Progress in
Medical Radiation
Physics

Volume 2

Progress in Medical Radiation Physics

Series Editor:

Progress in Medical Radiation Physics

Volume 2

Edited by
COLIN G. ORTON

Wayne State University School of Medicine,
Harper–Grace Hospitals
Detroit, Michigan

PLENUM PRESS · NEW YORK AND LONDON

ISBN-13: 978-1-4612-9458-0   e-ISBN-13: 978-1-4613-2387-7
DOI: 10.1007/978-1-4613-2387-7

© 1985 Plenum Press, New York
Softcover reprint of the hardcover 1st edition 1985
A Division of Plenum Publishing Corporation
233 Spring Street, New York, N.Y. 10013

Roy E. Ellis

This book is dedicated to the memory of our dear friend and colleague Professor Roy E. Ellis, who, until his untimely death, was an active member of our Editorial Board.

Throughout his career, Professor Ellis devoted himself tirelessly to the safe use of radiation for the welfare of mankind. He was a very practical man. He realized that radiation, used constructively, could be of considerable benefit, yet, if applied indiscriminately, was a hazard which needed to be controlled. He was a great leader both in the clinical application of radiation and in protection against its harmful effects. Those of us who had the pleasure of working with him will always relish the experience. It was an honor and a privilege to have had Roy Ellis as a member of our Editorial Board. His enthusiasm and vitality will be sorely missed.

Colin G. Orton

Contributors

J. A. Brace, Radiotherapy Department, Royal Free Hospital, London, England

Lewis Burkinshaw, Department of Medical Physics, University of Leeds, The General Infirmary, Leeds, England

T. J. Davy, Medical Physics Department, Royal Free Hospital, Hampstead, London, England

Kunio Doi, Kurt Rossmann Laboratories for Radiologic Image Research, Department of Radiology, University of Chicago, Chicago, Illinois

Paul A. Feller, Department of Radiology, The Jewish Hospital, Cincinnati, Ohio

W. A. Jennings, National Physical Laboratory, Teddington, England

James G. Kereiakes, Department of Radiology, University of Cincinnati College of Medicine, Cincinnati, Ohio

Stephen R. Thomas, Department of Radiology, University of Cincinnati College of Medicine, Cincinnati, Ohio

Preface

The *Progress in Medical Radiation Physics* series presents in-depth reviews of many of the significant developments resulting from the application of physics to medicine. This series is intended to span the gap between research papers published in scientific journals, which tend to lack details, and complete textbooks or theses, which are usually far more detailed than necessary to provide a working knowledge of the subject.

Each chapter in this series is designed to provide just enough information to enable readers to both fully understand the development described and apply the technique or concept, if they so desire. Thorough references are provided for those who wish to consider the original literature. In this way, it is hoped that the *Progress in Medical Radiation Physics* series will be a catalyst encouraging medical physicists to apply new techniques and developments in their daily practice.

Colin G. Orton

Contents

1-II. Physical Aspects of Conformation Therapy Using Computer-Controlled Tracking Units
T. J. Davy

1-III. Computer Systems for the Control of Teletherapy Units
J. A. Brace

2. Measurement of Human Body Composition *in Vivo*
Lewis Burkinshaw

3. Medical Applications of Elemental Analysis Using Fluorescence Techniques
Paul A. Feller, James G. Kereiakes, and Stephen R. Thomas

4. Basic Imaging Properties of Radiographic Systems and Their Measurement
Kunio Doi

Progress in
Medical Radiation
Physics

1-1

The Tracking Cobalt Project
From Moving-Beam Therapy to Three-Dimensional Programmed Irradiation.

W. A. JENNINGS

"My object all sublime
I shall achieve in time,
to let the punishment fit the crime,
the punishment fit the crime"

("The Mikado," Gilbert and Sullivan, 1885)

1. INTRODUCTION

In radiotherapy, the dose distribution achieved over the tumor volume, whatever its shape and location, and avoiding damage to surrounding tissues are vital components in the success or failure of the treatment. Moreover, by minimizing the volume irradiated, it becomes possible to deliver higher doses, and as first demonstrated by Ralston Paterson in Manchester, the higher doses tolerated lead to higher cure rates. However, adopting smaller volumes and higher dose levels call for higher precision in delivery, since Shukovsky,[1] Stewart and Jackson,[2] and Herring[3] have shown that there is a steep function relating dose with probability of cure and also risk of damage.

Thus, the present chapter is concerned with the approach to *conformation*

W. A. JENNINGS • National Physical Laboratory, Teddington TW11 0LW England.

therapy, a term used to describe treatments in which the high dose volume is shaped in three dimensions to conform to that of the target volume. In some instances, the target volume may be sinuous; for example, if one is to irradiate certain paths along which cells spread from a primary site, then the ratio in volumes between such sinuous tracks and "enclosing" cylinders may be considerable. Focusing on treating such sinuous tracks, Green[4] devised the *tracking* concept where the patient is tracked along in three dimensions, guiding the target path through an appropriate high-dose core produced by an "arc" therapy machine.

Conformation therapy has had to be approached step by step, and Part I of this contribution describes the development of the project to this end. References to parallel approaches undertaken elsewhere are made, but the present discussion is largely confined to work on this project carried out in London, initially at the Royal Northern Hospital (1945–1970), and later at the Royal Free Hospital (from 1970), to which the project was transferred.

A. Green, the radiotherapist in charge at the Royal Northern Hospital, had long believed in the dictum "let the punishment fit the crime!" In 1937, while working for the Medical Research Council, Green[5] had devised the first beam-directing calliper for use with a radium beam unit. This device was soon extended by Dobbie[6] to become the "back pointer" for X-ray therapy. Such devices constituted a significant advance in correctly aiming radiation beams at tumors. Multifield techniques were developed with a view to enclosing tumors in nearly-uniform dose zones, with the routine application aided by the use of jigs as well as back pointers.

The next step was to develop the concept of moving-beam therapy, in which either the patient or the beam moved during treatment while the beam remained pointed at the tumor. This approach had advantages with respect to both the percentage depth dose that could be achieved and the accuracy with which delivery could be made. The concept was not new; indeed, the idea was first suggested by Kohl in 1906, and a number of approaches were described in the literature prior to 1945. However, the whole approach was rigorously pursued by Green, Jennings, and Bush[7] at the Royal Northern Hospital during the late 1940s and early 1950s, and the hospital became a recognized center for the three principal types of moving-beam techniques, namely, rotating-chair therapy, arc therapy, and conical therapy, all with 250-kV X rays.

Generally, such techniques will produce nearly-spherical or nearly-cylindrical high-dose zones at a depth when using circular or cylindrical beam sections, respectively. By the later 1950s, B. S. Proimos in Athens and K. A. Wright in Boston had introduced synchronous-shielding and beam-shaping devices to moving-beam techniques to alter the shape of the high-dose zone and to protect radiosensitive organs. These techniques are summarized later.

At the Royal Northern Hospital, the 1950s saw the development by Green, Jennings, and Christie[8] of the tracking technique with a view to irradiating the spread of malignant disease from the primary growth along the lymph node chains, the object being to confine the high-dose zone to the sinuous track. This was achieved by motorizing the patient couch in order to "feed" the path to be treated through the high-dose zone produced by arc therapy with a narrow beam. The vertical and lateral movements of the machine, Mark I, were controlled by four quadrant-shaped electrical contacts around a metal rod bent to the shape of the track, and the existing 250-kV X-ray unit was used as the radiation source.

From the experience gained in treating patients with this machine, a new model was designed by A. Green and W. A. Jennings in collaboration with G. R. Fulker of TEM Instruments. Couch movements were to be controlled by a photo-electric profile-follower system, linked to a cobalt-60 unit. In order to fund what became known as the Tracking Cobalt Project, a public appeal for some £50,000 was launched in 1963 by Sir Bracewell Smith, a former lord mayor of London.

As part of thoroughgoing consultations, a study tour of the United States was undertaken, and in consequence, the need for additional features in the system became apparent. In particular, the high-dose zone within the patient cross section, hitherto nearly circular, was elaborated to fit elliptical shapes with variable dimensions, eccentricities and orientations as one progressed along the track. Achieving such distributions meant incorporating additional control parameters, including the means of altering the beam width and the angular velocity at set positions within each arc.

Following computations and experimental work with a cobalt-60 unit at St. Bartholomew's Hospital, the design of the Mark II machine was finalized (Jennings,[9] and Green[10]), and the necessary funding raised. The machine was completed, and due to a reorganization in the radiotherapy services for North London, it was installed at the Royal Free Hospital in 1970. The radiotherapist in charge was B. L. S. Skeggs, and a thoroughgoing dosimetry program was undertaken by Davy et al.[11] prior to commencing treatments.

However, the sheer complexity of both the treatment planning and the machine's control system proved excessive in routine use. Fortunately, the advent of the computer made possible the development of the third model during the evolution of this project. The new machine, Mark III, employing a new cobalt-60 unit, was designed by TEM Instruments according to the Royal Free Hospital's requirements, which were based on the experience gained with the analogue-controlled unit (Mark II). J. A. Brace designed and wrote the software to control and test the treatment unit (Brace et al.[12]). This new machine was entirely computer-controlled, not only for the tracking technique, but also for increased efficiency in repeat settings for multifield techniques and with respect to safety aspects through built-in verification systems. In parallel, the computer made possible significant advances in treatment planning, which entails multisection calculations and summations.

Parts II and III of this chapter discuss the current position with respect to the project at the Royal Free Hospital, a project that has thus evolved into conformation therapy by means of computer-controlled dynamic irradiation. Both the physical aspects of the treatment (Part II), and the computer machine-control systems (Part III), are considered. It is evident that the principles described could be applied to other types of radiation and equipment, but the present discussion is oriented to computer systems for the dynamic control of teletherapy units, and to treatment with photons, including the use of linear accelerators as well as cobalt-60 units. With the introduction of microprocessors, it is becoming possible to make this approach practicable on a wider scale, the sheer complexity involved having delayed its introduction in the past.

2. ESTABLISHING MOVING-BEAM TECHNIQUES AT THE ROYAL NORTHERN HOSPITAL, 1945–1955

2.1. Alternative Moving-Beam Techniques

With 250-kV X rays, delivering an adequate dose to a deep-seated tumor with minimal skin damage generally necessitates the convergence of several radiation beams at the tumor site. This can be achieved by a "cross-fire" arrangement, employing a series of separate tube settings. However, as early as 1906, Kohl saw the advantages of a continuously moving system in which an inclined X-ray tube revolved around a vertical axis through a supine patient's tumor, but this presented difficulties with sliding contacts for the high tension (HT) supplies. Alternative systems were suggested by Pohl in 1913, and first described by Meyer in the same year, in which either the patient sat on a rotating chair with a stationary beam aimed horizontally or the patient lay still while the tube oscillated in an arc overhead.

In the 1930s and early 1940s, these techniques were considered by a number of authors, mostly in Germany, France, Denmark, England, the United States, and Japan (see Ref. 7). However, a high proportion of the published papers were purely theoretical in character, with little or no discussion of dosimetry, patient-positioning techniques, or clinical results. During the period from 1945 to 1955, all three of the principal moving-beam approaches were developed at the Royal Northern Hospital in London into practical techniques and put into routine use. In consequence, the center became a focus of attention for moving-beam radiotherapy at that time, even though restricted to 250-kV X rays.

2.2. Conical-Rotation Therapy

In this technique, first described in 1932 by Knox and Caulfield,[13] and later by Henschke,[14] the X-ray tube is set at an inclined position over the patient while the latter is rotated on a treatment couch around an axis passing through the tumor. This approach overcomes the HT supply problem that arose in Kohl's original proposal for revolving the inclined tube itself above a stationary patient. This technique, termed conical-rotation therapy, has the major advantage over "arc" therapy (see Section 2.4) that it can be used with any X-ray installation provided the clearance under the tube is adequate.

This technique was taken up in 1945 at the Royal Northern Hospital and developed for treating deep-seated tumors by irradiation from both sides of the patient or for treating shallow tumors from one side only, the latter entailing the use of wedge filters. A full description of the technique was published in 1949 (Green et al.[7]). Figure 1 indicates some of the salient features. It should be noted that using a plastic (Perspex/Lucite) disc as shown served a number of purposes: It provided compression and assisted in localization and beam positioning. Further, the disc and cone enabled one to apply dose distributions, measured in tissue-equivalent wax phantoms, directly to the same geometrical shapes in treatment if allowance for heterogeneity were made when appropriate.

Sites treated included the uterine cervix, the body of uterus, ovary,

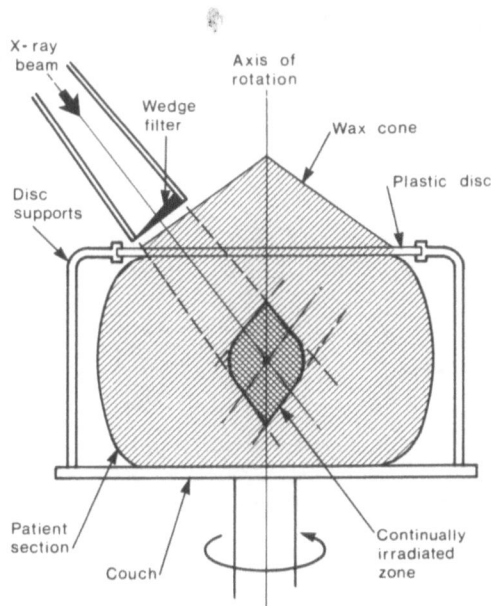

Figure 1. Conical rotation therapy
(Green, Jennings, and Bush, 1949).

bladder, rectum, and bronchus. It was just possible to reach the body's extremities, and hence also treatment of the thyroid, tongue, and brain. For treatment of the thorax, or extremities, special padded plastic shells and (evacuated) bolus bags were also used. After several years' use, this treatment was found to be an accurate and precise method of delivering high tumor doses, and the constitutional effects for a given tumor dose were less than in other forms of treatment.

2.3. Rotating-Chair Therapy

In this technique, first described by H. Meyer in 1913, the patient sits in a special chair, which rotates around an axis passing through the tumor while a stationary horizontal X-ray beam is aimed at the tumor, as indicated in Figure 2. This approach has been discussed by a number of authors over the years (see Ref. 7) and in particular by Nielsen and Jensen[16] in 1942, who employed screening during treatment to ensure correct positioning.

In practice, this technique is limited to treating tumors in the thorax, such as bronchus or esophagus, and entails using specially designed chairs with immobilizing straps and generally with devices to hold the arms above the radiation beam. At the Royal Northern Hospital, a special chair was constructed to be mounted on top of the rotating couch used for conical therapy, the chair incorporating lateral adjustments in order to place the axis of rotation through the tumor.

This technique permits achieving a high percentage depth dose, but the dosimetry is more complex. Individual patient contours have to be taken and beam dose contributions summated at a series of angles to simulate rotation. The effects of heterogeneities, and the chair pillar, will require corrections for

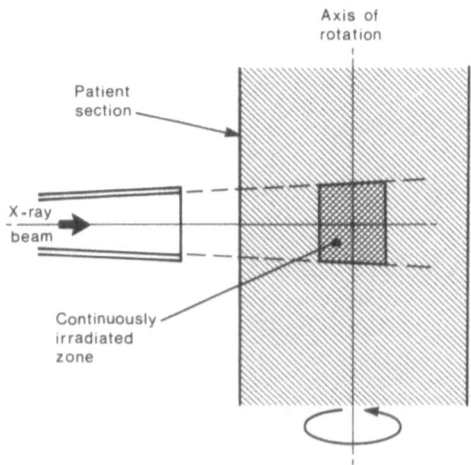

Figure 2.   Rotating-chair therapy.

both dose rates and possible displacement of the maximum dose position from the axis. Phantom measurement programs were carried through, using different sized body and lung cross sections. The body was simulated by tissue-equivalent material ("lincolnshire bolus") and the lungs by cork granules or by the breakfast cereal Grapenuts, with a density of ~0.3. One such phantom is shown in Figure 3(a), with a series of measured dose distributions in Figure 3(b).

2.4. Arc or Pendulum Therapy

The third principle type of moving-beam therapy entails the use of an X-ray tube oscillating to and fro in an arc above the stationary patient. The technique was also first described by Meyer,[15] and discussed by a number of authors over the years (see Ref. 7). Unlike conical therapy, this approach calls for a more complex X-ray installation.

To establish the technique at the Royal Northern Hospital, the existing gantry of a 250-kV Marconi X-ray unit was modified by adding alternative heart-shaped cams, which while rotating continuously, provided for tube oscillations of 120°, 150°, or 180° around a preset angular position, as illustrated in Figure 4. In order to use a beam diaphragm as close to the patient as practicable, a special couch was constructed with removable side panels, with the separate couch used for conical and chair therapy rolled out of the way on inset floor rails during arc treatments.

This technique is illustrated in Figure 5, where attention is drawn to the displacement of the high-dose area from the continuously irradiated zone around the oscillation axis if only a 180° arc is applied. This displacement will be rectified if both sides of the patient are treated in turn. If only one side is treated, for a shallow tumor, it will be necessary to aim at an axis deep to the tumor center. The dosimetry is essentially similar to that mentioned for the rotating-chair technique in Section 2.3 except that it tends to be more complex, since arc angles of under 360° are more likely to arise, and displacement problems are accentuated.

In contrast to the use of isodose charts for multibeam techniques at fixed source–skin distances, using rotation techniques in chair and arc therapy called for new depth dose data based on fixed source–tumor distances but with varying source–skin distances. To this end, new tumor–air ratio charts were developed, this ratio being that for the dose at the depth with and without the patient present, for a given arc radius at a series of angular settings.

From the clinical standpoint, arc therapy proved the most versatile in application. Special localization and setting-up procedures were devised to ensure accuracy and repeatability. Again, patients were found to tolerate such treatments well, and higher doses to smaller volumes became possible.

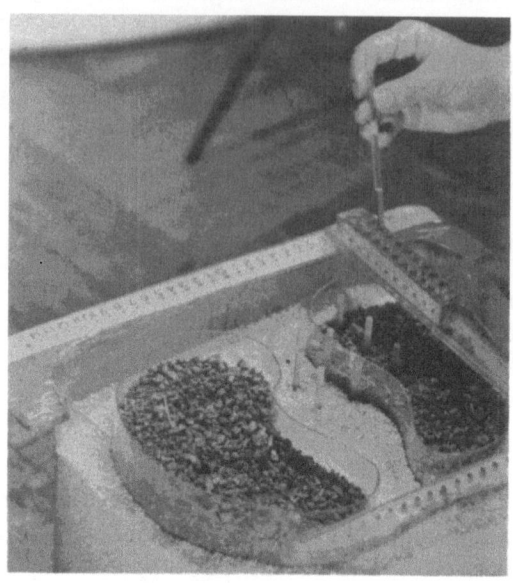

a

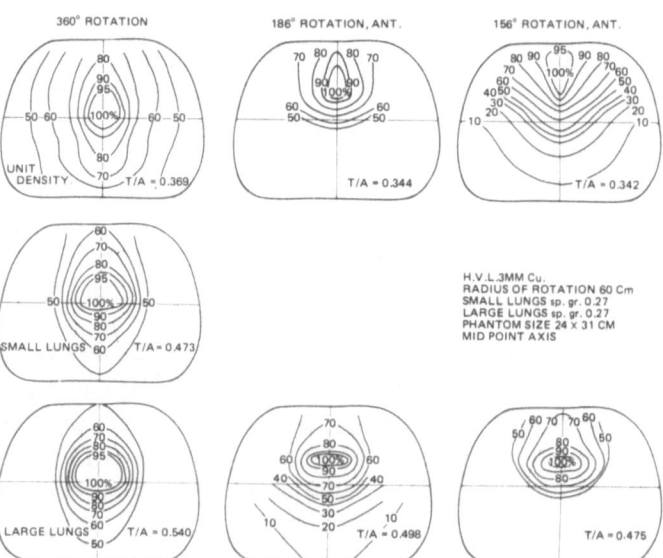

b

Figure 3. (a) Phantom for dose-distribution measurements. Microionization chambers inserted in a 2 × 2 cm grid within a "lincolnshire bolus" body with cork granules for lungs. (b) Measured dose distributions to illustrate the effect of changing the lung size and angle of rotation.

Figure 4. Modified 250-kV Marconi X-ray unit for arc therapy, using heart-shaped rotating cams. A wax phantom is shown for distribution measurements using microionization chambers.

2.5. Relative Percentage Depth Doses Achieved

Although the full dosage distribution obtained is necessary for comparing the merits of the moving-beam techniques, the percentage depth dose attainable in each instance is a significant factor. Moreover, the use of more penetrating X rays would clearly be advantageous. To this end, two studies were carried out by the author from 1955 to 1956 at the Argonne Cancer Research Hospital in Chicago. An experimental program was undertaken to establish conical rotation with a 2-MV Van de Graaff generator (Jennings and

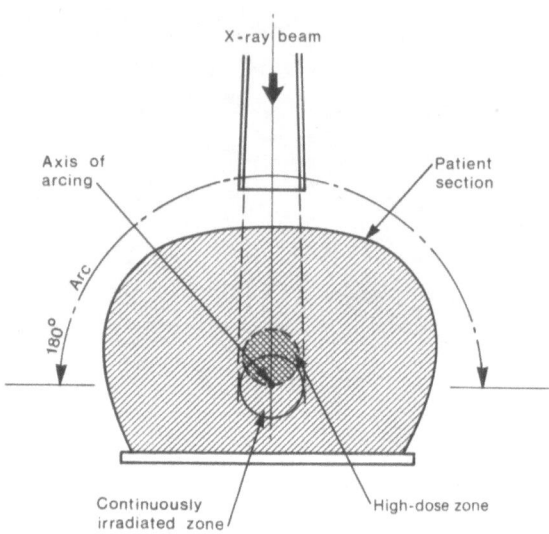

Figure 5. Arc therapy.

McCrea[17]), and a theoretical study was made of the percentage depth doses achieved by conical therapy at 250-kV versus 2-MV X rays, and by conical versus chair and arc therapy at 250 kV under widely varying conditions (Jennings[18]). These papers should be consulted for a detailed analysis.

3. THE TRACKING CONCEPT AND THE OPERATION OF THE MARK I TRACKING MACHINE, 1957–1959

3.1. The Spread of Malignant Disease

The common pathological spread of malignant disease in moderately early cases involves lymph node tracks, for example, in carcinoma of the cervix and seminoma of the testis. The lymphatic spread in the case of the former commonly involves lymph nodes from the obturator region below to the para-aortic nodes above, on the affected side. As pointed out by Green,[4] such a series of nodes lie on a narrow undulating pathway, as illustrated in Figure 6, and hence, the objective should be to confine the high-dose region to such a sinuous shape. Such a step should lead to improved cure rates on two counts: (1) results achieved in the treatment of smaller volume to higher dose levels (see Refs. 1–3) and (2) the practical irradiation of lines of lymphatic spread in appropriate cases.

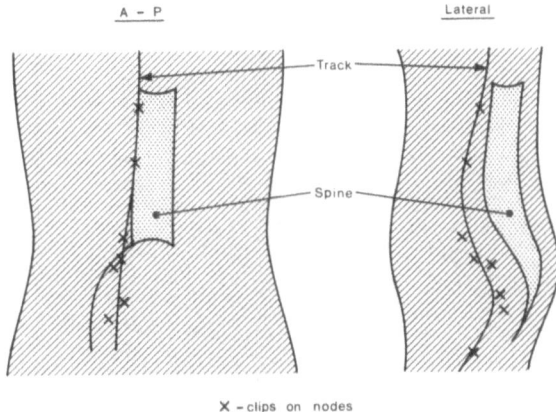

Figure 6. Lymphatic spread in carcinoma of the cervix. Crosses indicate location of metal clips applied to sites of nodes commonly involved (Green, 1959).

3.2. Realization of the Tracking Principle

In order to confine the high-dose zone to a sinuous track, it is necessary either to develop a system of variable-beam diaphragms or to "track" the patient along in three dimensions. The latter technique was introduced at the Royal Northern Hospital in 1957 (Green[4]).

Firstly, a small-volume/high-dose zone could be achieved at a depth by means of the 250-kV arc-therapy unit already in service at the hospital. A series of arc dose distributions were available, indicating the extent and location of the high-dose zone for various beam depths and arcing angles.

Secondly, in order to confine the high-dose zone to the sinuous track, the patient would have to be moved both laterally and vertically while tracking along, as indicated in Figure 7. A comparison between the volumes of a straight cylinder containing the track and the "sinuous" cylinder surrounding the track indicates the significant advantage to be achieved at any rate in principle.

In order to achieve the necessary patient movements in three dimensions, the existing arc-therapy couch was first motorized in collaboration with Sierex Ltd. This entailed the addition of three motors to control the couch's motion in the x, y, and z directions, the x direction being along rails on the floor. The y and z movements were controlled by an electromechanical system, as indicated in Figure 8. A length of brass rod, bent into the shape of the required three-dimensional path, was attached to the couch. A ring with four quadrant contacts was placed over the rod and clamped to a fixed floor stand. Then, while the couch tracked along in the x direction, the ring contacts activated the y and z movement motors during contact, so that the

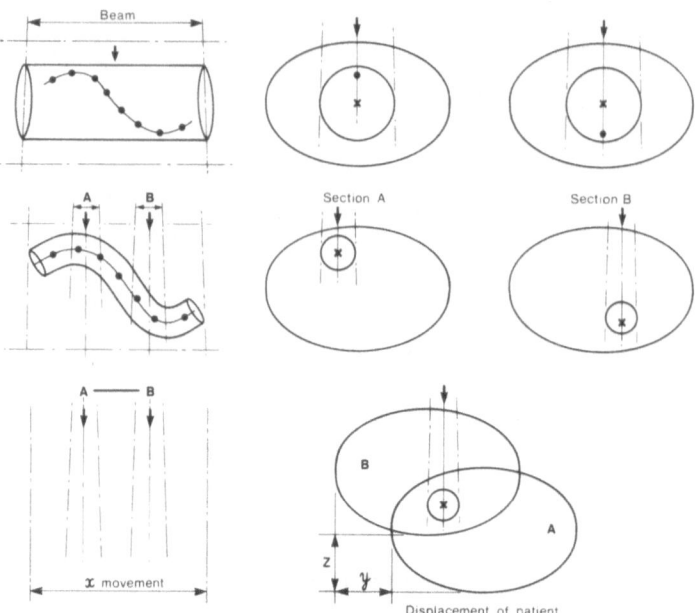

Figure 7. Irradiation of the sinuous track. Diagram illustrates reduction in volume needed to enclose such a track by the lateral and vertical displacement of the patient while being "tracked" along.

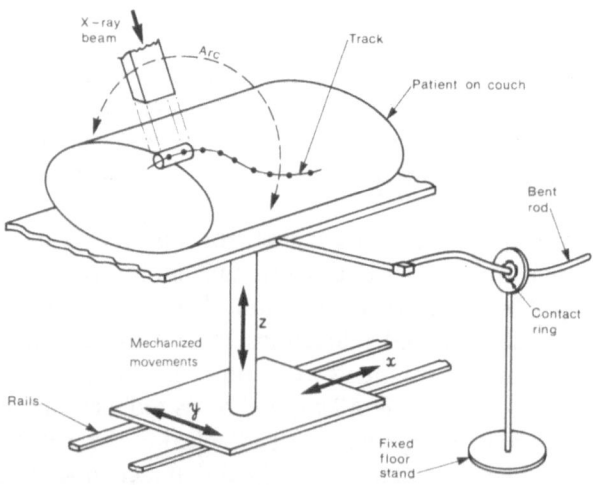

Figure 8. The "tracking" principle, using a metal rod bent to the shape of the track.

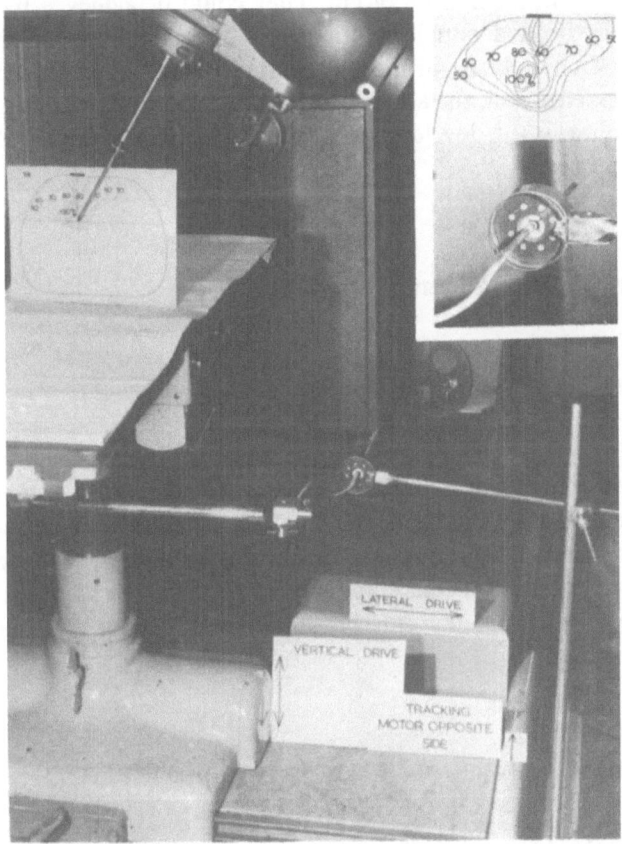

Figure 9. The "tracking" couch, illustrating the bent metal rod passing through the ring of contacts. The dose distribution indicates the displacement of the high-dose zone from the axis of arcing (Green, 1959).

rod passed through the fixed ring, and hence, the high-dose zone followed a parallel track in the patient. Figure 9 illustrates the equipment itself. In order to bend the rod to the correct shape, its projection had to be correct in both the y and z planes, as indicated by lines drawn on lateral and AP radiographs, respectively, allowing for magnification.

Before describing the achievement of uniform dosage along such a relatively simple track, one with a nearly circular cross section, it is worthy of note that even in 1959, Green envisaged variations in the cross section of the high-dose zone as one proceeded along the tack. Thus, in his paper to the Royal Society of Medicine, Green[4] included two proposed oval-shaped dose contours of narrow depth for treating para-aortic nodes just in front of the kidneys and behind the mass of gut. They could be achieved by combining two 80° arcs from a cobalt-60 unit or better still from a 4-MeV linear accelerator. Generally, he proposed the application of appropriate *predetermined shapes*. However, the application of these concepts, even for a cobalt-60 unit, had to await the development of the Mark II machine, as discussed in Sections 4.5 and 5.

3.3. Achieving Uniform Dosage along the Track

3.3.1. INTRODUCTION

It will be appreciated that when the track lies at a greater depth, the dose rate will be less than when the track is shallower, and hence, the track will have to spend more time in the beam to receive the same dose. Other problems include the dosage falling off at the end of the track. In consequence, a full computational and measurement program, including the use of small ionization chambers and film emulsions in phantoms, was undertaken.

Briefly, the dosimetry can be summarized under four headings as follows.

3.3.2. DERIVATION OF THE DOSAGE DISTRIBUTION AT EACH OF A SERIES OF SECTIONS AT SELECTED INTERVALS ALONG THE TRACK

One has to know (1) the axial and maximal dose rates, (2) the displacement if any in the y and z directions of the maximum dose point from the track axis, and (3) the coverage of at least the 80% and 50% dose contours. An example is shown in the inset in Figure 9.

The sectional dosimetry is simply an extension of that needed for arc therapy. The displacement of the maximum dose point from the beam axis will need to be taken into account in the setting-up procedures.

3.3.3. MAINTAINING THE DOSE TO THE ENDS OF THE TRACK COMBINED WITH A SHARP FALL-OFF BEYOND THEM

The solution adopted is shown in Figure 10, which entailed using lead rubber at the track ends and reducing the tracking speed at the ends in order to compensate for the reduction in field width when crossing the rubber.

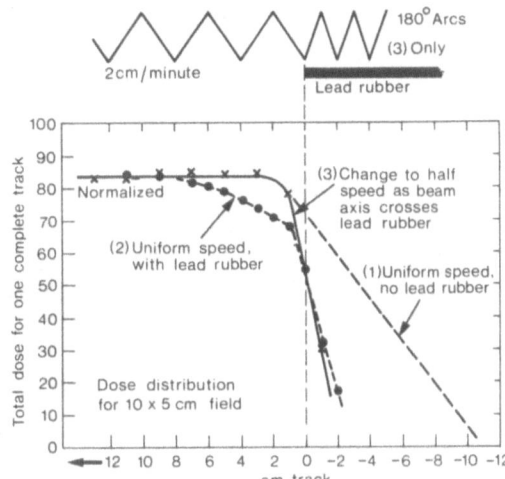

Figure 10. Maintaining the dose to the ends of the track, combined with a sharp decrease beyond them. Example of measured data.

3.3.4. COMPENSATION BY VARIATIONS IN TRACKING SPEEDS FOR VARIATIONS IN SECTIONAL DOSE RATES ALONG THE TRACK

By tracking at a series of speeds that are inversely proportional to the maximum sectional dose rates, the integrated dose per unit time remains approximately constant. The dose distribution, measured in a phantom, along a track of increasing depth is shown in Figure 11.

3.3.5. RELATION OF THE SECTIONAL DOSE RATE TO THE DOSE PER TRACK

This problem includes (1) variation in the dose rate across the main beam in the direction of travel (x), yielding a mean effective figure of $\sim 90\%$ of the axial dose rate, independent of depth, providing the track length is at least double the field length; (2) the contribution of scattered radiation outside the beam, which depend on the track length and depth, and field size. Experimental data was acquired for these factors, as illustrated in Figure 12.

3.4. Practical Application of the Tracking Technique

Taking seminoma of the testis as an example, arteriographs were taken and used to determine the position of the nodes, and then the track dimensions and settings were determined. Figure 13 illustrates in a single combined chart the essential physical data required for "programming" the patient's treatment. This chart consists of (1) the positions, marked A, B, C, D, and E, at which the patient's shape was measured and drawn for sectional

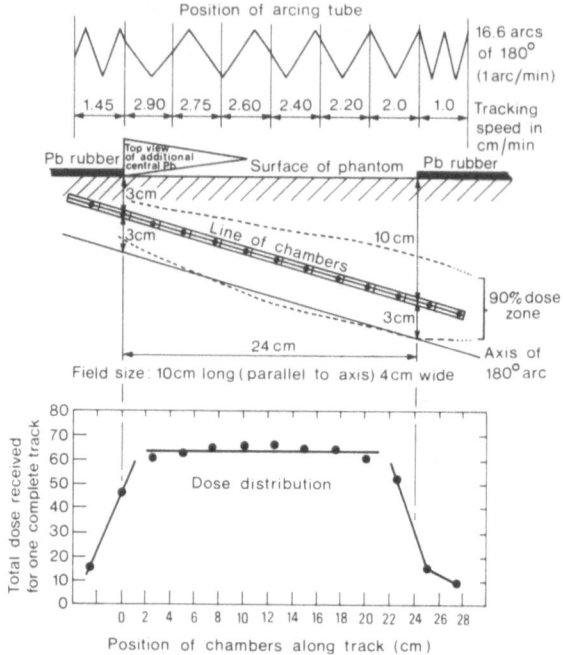

Figure 11. Compensation, by variation in tracking speed, for variations in sectional dose rates along the track. Example of measured data.

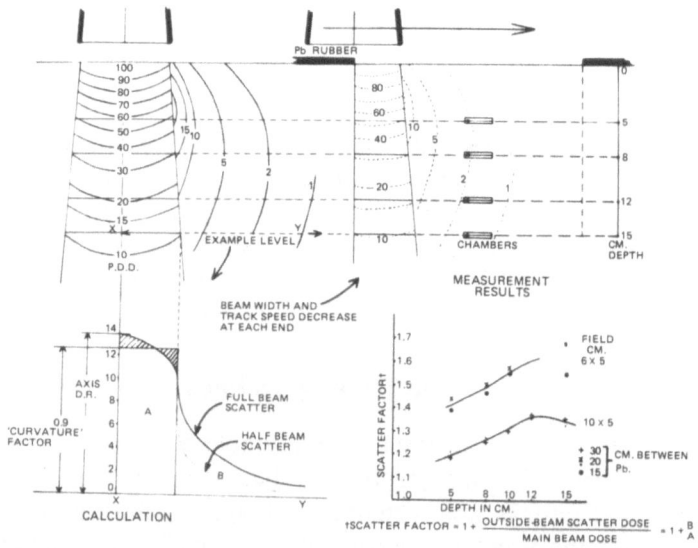

Figure 12. Derivation of the dose per track at a point on the axis. Example of calculated and measured data.

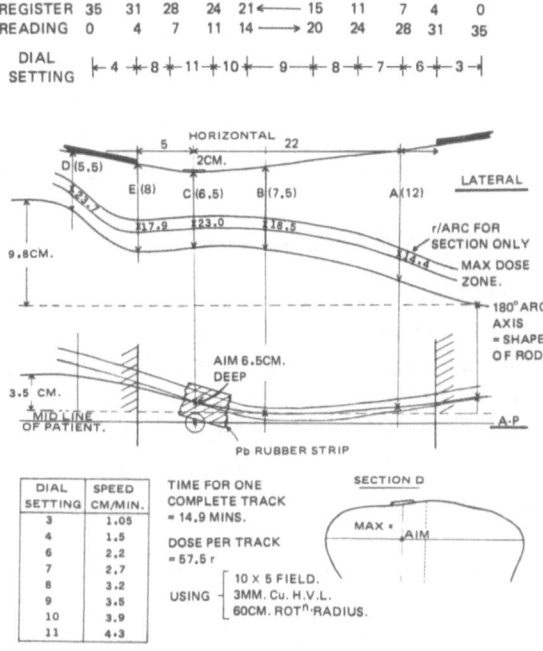

Figure 13. Program chart for the practical application of a particular tracking treatment (seminoma of the testis).

calculations; (2) the maximum dose rates applicable to each section alone; (3) the tracking speed and corresponding dial numbers, selected to compensate for variation in the sectional dose rates as calculated in 1; (4) the register readings at which changes in dial (i.e., speed) numbers are made either manually or preset (automatically); (5) essential reference points and measurements required for setting up the patient in the treatment position.

A series of patients were treated with this equipment, later referred to as the *Mark I tracking unit* in view of its successors (see Section 4). The project was presented at the Ninth International Congress of Radiology in Munich in 1959 (Green, Jennings, and Christie[8]).

4. STEPS TOWARD AN IMPROVED TRACKING MACHINE, 1960–1965

4.1. The Need for Penetrating Radiation

Based on the experience gained with the Mark I model, a new machine was planned by Green and Jennings. It had long been envisaged that a higher energy radiation source would be used in order to achieve higher percentage depth dose distributions, and although a linear accelerator would provide minimal penumbra, a cobalt-60 gamma ray unit with a large source was adopted as the next step, in view of its advantage in terms of reliability and cost at the time. TEM Instruments was approached with a view to adapting a forthcoming model, a Mobaltron 100, to the tracking technique.

4.2. Developing an Improved Control System

G. R. Fulker of TEM Instruments devised a "profile-follower" technique, employing photo-electric cells, which operated servo motors to control the various parameters. Thus, to control the y and z movements of the couch, the former bent metal rod was replaced by two profiles, the black on white edges corresponding to projections of the rod in the two planes. These projections are, indeed, the information provided by AP and lateral radiographs directly if account is taken of the enlargement factor. Figure 14 illustrates the principles of the system.

Similar profiles, also controlling servo motors, could be used to control the tracking speed in the x direction and, by motorizing the diaphragm system, both the y and z field dimensions. The latter variables were needed to reduce the field length at the ends of the track (to maintain the dose at the ends, as before) and to vary the track width if necessary as one progressed along the track.

In order to apply the system, the preceding five control profiles would be

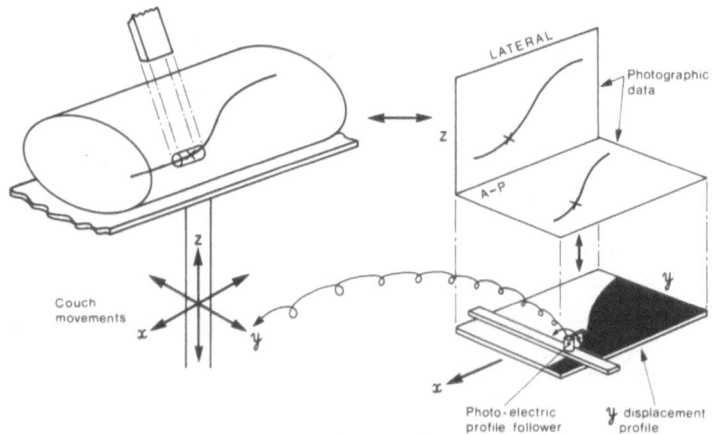

Figure 14. Application of the tracking principle by using profile followers. The profiles are derived from the patient's radiographs, and the couch movements are controlled by servo motors.

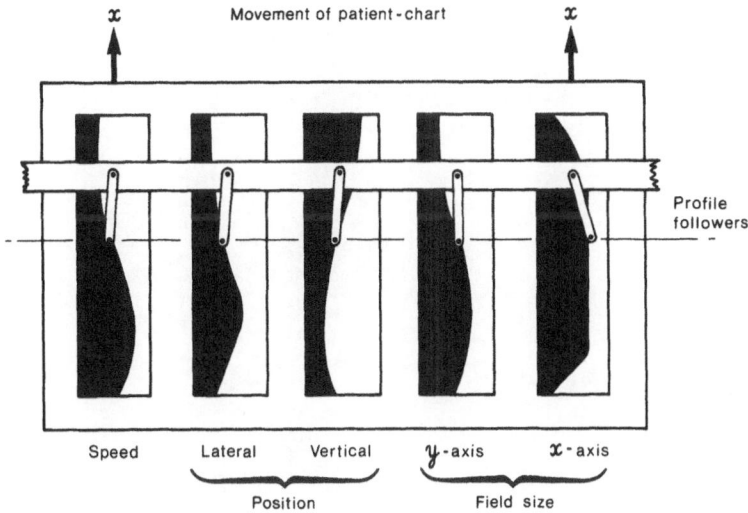

Figure 15. Diagram to illustrate a patient chart containing five control profiles for the parameters indicated (see also Figure 22).

prepared and mounted as a single patient chart, as indicated in Figure 15, and the chart itself would progress in the machine console at the same rate as the couch progressed in the x direction.

4.3. An Appeal for Funds for the Tracking Cobalt Project

By 1963, it had become clear that the realization of the Mark II machine depended on a major injection of funds, and a Public Appeal was duly launched for some £50,000 with the enthusiastic support of a committee of prominent people chaired by Sir Bracewell Smith, a former lord mayor of London. The vice chairman was Viscount Bearsted, and other members included Sir Ralph Richardson, the actor, and J. R. McNeill Love, an eminent surgeon. Numerous fund-raising functions were held, including an all star Sunday charity show at a London theatre, a major reception with the Lord Mayor of London, the sheriffs of the city of London and the mayors and mayoresses of all the metropolitan boroughs of London,* and a large-scale darts tournament (for the Bray Cup) in the pubs of North London. Perhaps the peak event was a major carnival in the streets of Islington, the London borough where the Royal Northern Hospital was located (see Figure 16). The necessary funds came in at a steady pace, and work on an improved machine continued.

* This was at 8 p.m. on 22 November 1963, and was sadly interrupted by an announcement of President Kennedy's assassination.

Figure 16. Fund raising for the Tracking Cobalt Project (1964).

4.4. Consultations and Study-Tour

Numerous consultations with experts within the United Kingdom were held, and it was also decided to undertake an extensive study-tour in the United States, visiting various centers of excellence, and in particular the Leahy Clinic, where alternative conformation therapy techniques were known to be under development (see Section 6).

4.4.1. TECHNIQUE AND DOSIMETRY FACETS

The centers in the United States visited in 1964 by the author are listed in Table 1. In 12 of these 14 centers, a symposium was held regarding the project in order to invite comment. Various types of moving-beam techniques were seen and studied in some departments (centers 4, 5, 6, and 9); dosimetry techniques using digital computers (centers 1, 5, and 6) and analogue computers (centers 6, 7, and 9) were also studied. Measurement techniques and phantoms were discussed; in particular, the new Alderson Rando phantom developed at center 6 (Alderson et al.[19]) was of direct

Table 1. Study-Tour, U.S.A., April–May 1964

Center	City
1. Memorial Hospital and Sloan-Kettering Institute[a]	New York
2. Grace–New Haven Hospital[a] and Yale University	New Haven
3. Massachusetts General Hospital[a]	Boston
4. Leahy Clinic—Massachusetts Institute of Technology[a]	Boston
5. Princess Margaret Hospital and Ontario Cancer Institute[a]	Toronto
6. University of Chicago clinics and Argonne Cancer Research Hospital[a]	Chicago
7. University of Illinois Medical Center[a]	Chicago
8. Michael Reese Hospital	Chicago
9. Veterans Administration Research Hospital[a]	Chicago
10. University of Pennsylvania Medical Center[a]	Philadelphia
11. University of Maryland Medical Center[a]	Baltimore
12. Johns Hopkins University Hospital[a]	Baltimore
13. University of North Carolina Medical Center } [a] joint	Chapel Hill
14. Duke University Hospital }	Durham

[a] Denotes an address given (by W.A.J.) for a prearranged symposium on the subject of "tracking."

interest, and such a phantom was subsequently acquired for the project. From the diagnostic standpoint, the current development of scanning isotope machines at several centers (1, 3, 5, 6, 10, and 12) was of immediate interest also.

4.4.2. CLINICAL FACETS

While the author studied the technical and dosimetry facets of work related to the project, Green engaged in wide-ranging consultations at these centers with a view to determining the lines of spread of malignant cells from primary growths, along lymphatic pathways, in relation to the possibility of irradiating such tracks with accuracy. Indeed, his travels were even more extensive, encompassing additional centers in Washington, Detroit, Houston, San Francisco, and Seattle, together with several cities in Canada and a separate visit to Scandinavia.

Green's approach entailed requests to senior radiotherapist colleagues for recommended contacts with surgeons possessing vast experience in particular diseases who were known to be keen observers at operations. Cross-questioning such nominated surgeons led to surprising uniformity in information on specific situations in the body, confirming that the lymphatic spread along normal anatomical channels was, indeed, generally predictable.

Such an approach was more valuable than postmortem evidence, when the disease will be more widespread. It was agreed that the results of both surgically extirpating nodes involved in squamous carcinoma, and also large-field radiotherapy, were poor. It was also agreed that the high localized dose achievable by the "tracking-cobalt" technique should prove effective, at the same time minimizing radiation complications and damage to surrounding normal tissues. Lymphangiography and vascular studies were also seen and studied at many centers, and the radiotherapists were satisfied that they would serve to localize the lymph nodes with sufficient accuracy. Indeed, the persistence of contrast medium in lymphatics is a direct help in precisely placing beams in the tracking technique. Work done at the Royal Northern Hospital suggested that in some instances, lymph node tracks may be related to bony structures with a reasonable degree of accuracy, irrespective of variations in superficial fatty tissue, which would make localization simpler (Green[10]).

4.5. The Need for Elliptical Dose Contours

For carcinoma of the cervix, the position of the nodes that may require treatment can be located on a series of anatomical cross sections, taken at intervals along the line of spread. Figure 17 illustrates the high-dose zones needed at three positions along the track to be irradiated, and it became apparent that the former technique producing essentially circular dose contours at a section would not be optimal. As a consequence of this study, it was decided to aim at producing elliptical dose contours of varying

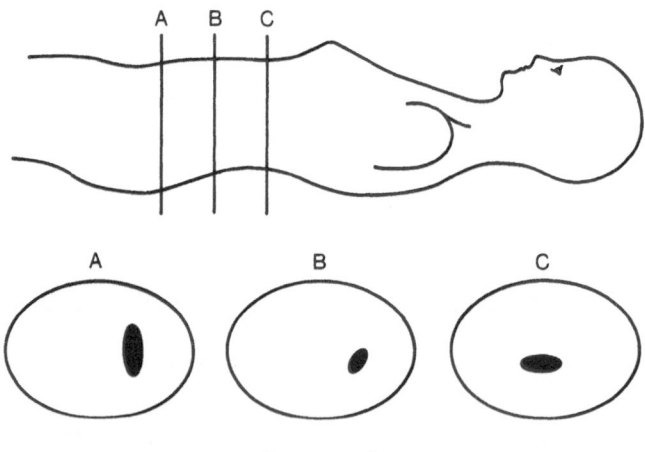

Cross sections

Figure 17. The need for elliptical dose contours of varying dimensions, eccentricities, and orientations (spread from carcinoma of the cervix).

dimensions, which would entail introducing additional machine-control parameters.

5. DOSIMETRY AND TREATMENT PLANNING FOR THE MARK II TRACKING MACHINE

5.1. Achieving Elliptical Dose Contours—the Approach Adopted

By mid-1964, it had become clear that elliptical dose contours would have a significant advantage over circular contours as an objective in sectional distributions for the tracking technique. Moreover, such ellipses would have to have varying dimensions, eccentricities, and orientations, depending, for example, on the position of the nodes to be encompassed.

Such an objective could be achieved by further elaborating the machine design already under development. Thus, arc irradiation can produce an elliptical dose distribution by varying the field width and/or the angular velocity during the revolution. It was decided to alter both of these parameters simultaneously, at set positions during an arc, as illustrated in Figure 18. Further, in order to provide adequate flexibility with respect to the eccentricity and orientation of the dose contours, the angular positions where field width and angular velocity changed, and the position of the oscillation axis, also required preset facilities, as shown in Figure 19.

In order to render these additional parameters amenable to control by profiles and servo motors, as for the already incorporated parameters, Fulker

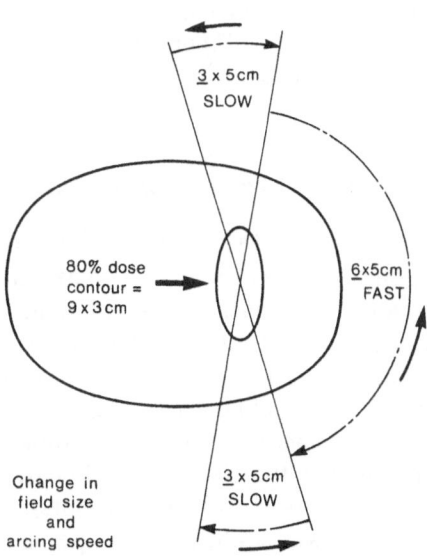

Figure 18. Achieving elliptical dose contours.

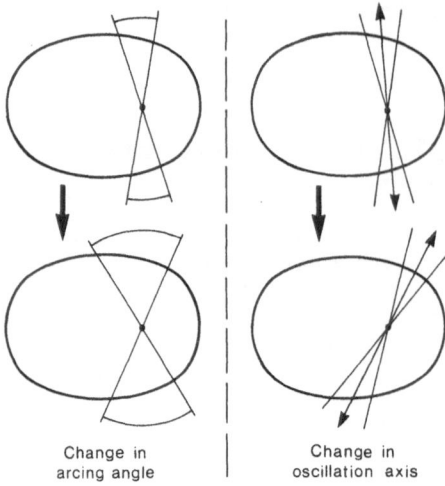

Change in arcing angle   Change in oscillation axis

Figure 19. Other parameters determining the eccentricity and orientation of elliptical dose contours (see also Figure 18).

(of TEM Instruments) adopted a "ratio" approach. Thus, the arcing speed ratio was a setting for the relative angular velocity in the respective sectors (i.e., "slow/fast" in Figure 18). The "arcing field size" ratio was the ratio in the respective field widths in the sectors (i.e., 3 and 6 cm in Figure 18), and the "arcing angle" ratio was the ratio of the sector angles. All these ratios could be varied from 1:1 to 1:4. With a setting for the orientation of the oscillation axis, and one for opening and closing the shutter, an additional five parameters were added to the five parameters formerly required, as shown in Figure 15 (Section 4.2). Thus, a second patient chart was needed, which would also progress in the machine console at the same rate as the couch progressed in the x direction. Figure 22 (Section 8.1) illustrates one of the two patient charts in the console.

5.2. Sectional Dose Computations

In view of the number of dose distributions, and the number of variables entailed, it was decided to make use of the then newly available Elliott Medical Automation radiation treatment planning service at University College Hospital, due to J. S. Clifton. Arc irradiations were simulated by summing a series of fields at 10° intervals over the required angles of oscillation. Changes in field size could be inserted and weighting factors used to allow for changes in arcing speed. A number of studies were carried out; for example, it was shown that the shape and dimensions of the 80% elliptical contour was relatively insensitive to its orientation, or the patient cross section, and its accuracy was confirmed by experimental data (see Section 5.3).

5.3. Sectional and Track Dose Measurements

An Alderson multisection Rando average-man phantom (Alderson *et al.*[19]) was purchased, and an experimental measurement program was conducted at two other centers in the United Kingdom, in view of the absence of a cobalt-60 machine at the Royal Northern Hospital.

Two approaches were adopted: one, using small ionization chambers; and the other, using film dosimetry, with both methods employing the new phantom. At St. Bartholomew's Hospital, with the collaboration of G. S. Innes, a cobalt-60 unit was used for both sectional-distribution studies with complex arc irradiations, and also for straight-line tracking, the treatment couch being motorized for the latter movement only. The ionization chambers were used to make comparisons with film dosimetry in a single plane, the latter facilitating multiplane measurements. A visit was also made to Norfolk and Norwich hospital with the collaboration of C. M. Pennington, to measure static beam and simple arc distributions, since the first Mobaltron 100 machine had recently been installed there. Dosimetry techniques relating to film dosimetry were studied at University College Hospital and the Royal Marsden Hospital.

5.4. Track Dosage

By selecting certain sites, such as carcinoma of the cervix, preliminary but detailed studies were undertaken to assess the dose distribution along the tracks. The tracks were divided into 2.5-cm lengths, and distributions were first derived for each cross section. The relative arc speeds needed to complete a sectional distribution had to be coordinated with the tracking speed, and the latter depended on the dose rate. The field length used was 5 cm, so that the sections tracked, and contributions from one section to the next, had to be estimated as well as the scatter from outside the primary beam. In practice, these effects tended to smooth out the distributions a little and called for film and ionization chamber measurements to confirm the estimations.

It will be appreciated that these studies were of a preliminary nature and a full measurement program would have to await the installation of the Mark II machine itself. Moreover, more comprehensive computational techniques would be needed for three-dimensional planning and, indeed, were soon to appear, particularly from Van de Geijn.[20,21] The dosimetry program following installation is summarized in Section 8.

The necessary funding continued to come in from the public appeal and the machine design was finalized, and the whole approach was presented to the First International Conference on Medical Physics in Harrogate and to the Eleventh International Congress of Radiology in Rome, both in 1965

(Jennings[9]). At the same time, Green made a short contribution to *Nature* explaining the principles of tracking radiotherapy (Green[10]).

6. ALTERNATIVE APPROACHES TO CONFORMATION THERAPY UNDER PARALLEL DEVELOPMENT ELSEWHERE—SYNCHRONOUS BEAM-SHAPING AND SHIELDING

The prime purpose of the present contribution is to discuss the development of the tracking technique, but no discussion relating to conformation therapy would be complete without some reference to the most significant parallel developments to this end in other centers. Owing to the sheer complexity of such programs, very few centers chose to engage in major efforts beyond the use of wedge filters and special arc techniques.

However, workers in Boston in particular were enthusiastically developing various devices, used synchronously with moving-beam techniques, for beam-shaping and shielding in the treatment of a series of sites. In the late 1950s, the team constituted K. A. Wright, B. S. Proimos, and J. G. Trump from the Massachusetts Institute of Technology (MIT), working in cooperation with M. I. Smedal, D. O. Johnson, and F. A. Salzman at the Lahey Clinic in Boston (Wright *et al.*[22]). Indeed, during the study tour undertaken by Jennings and Green in 1964, this center was visited and its work was discussed in detail with Wright and Trump at the clinic (Section 4.4). Moreover, one member in particular of the initial team, B. S. Proimos, who returned to his native Athens in 1961, developed a series of ingenious devices during the 1960s aimed at conformation therapy, which he described in a series of papers in the journal *Radiology* and then exhibited at the International Congresses of Radiology in Montreal in 1962 and Rome in 1965 (Proimos[23-28]).

Parallel work continued at the Leahy Clinic in Boston (Ilfeld *et al.*[29], and, indeed, in the late 1970s, Boston became a center for computer-controlled radiotherapy, this time at the Harvard Medical School, with a team comprising M. B. Levene, P. K. Kijewski, L. M. Chin, B. E. Bjarngard, and S. Hellman (see Section 10).

Returning to the work initiated at the Leahy Clinic in cooperation with MIT, the objective was twofold: firstly, to *shape the field* by interposing sufficient thicknesses of absorbant material within the beam cross section (in addition, the beam could be intensity modulated by filters to adjust for variations in the anatomical thickness); secondly, to *shield* certain radiation-sensitive regions within the field. Using rotation techniques, both objectives were accomplished by rotating the appropriate absorbers in synchronism with the rotation of the patient, by either mechanical linkage for horizontal beams or through gravity-orientation in arc techniques. Sites treated included the breast and chest wall with protection for the spinal cord, the thyroid with protection for the back of the neck, lymphomas with protection for the lungs, seminomas with protection for the kidneys, the head with protection for the eyes, and the uterus with protection for the bladder and rectum.

To illustrate these techniques, Figure 20(a) and 20(b), taken from one of the papers by Proimos,[25] illustrates a gravity-oriented shaping device, and Figure 21(a) and 21(b), taken from a later paper by Ilfeld et al.,[29] illustrates an elaborate set-up for irradiating advanced cervical carcinoma, including shielding devices.

A discussion of alternative approaches to conformation therapy in the 1960s must also include a reference to the work of S. Takahashi and his colleagues at the University of Nagoya in Japan. Takahashi[30,31] introduced the concept of variable *multileaf collimators* for use with a cobalt-60 rotational unit. The diaphragm system was controlled by multileaf cams within the cobalt unit head that could vary the field shape synchronously with the rotation. Thus, a sinuous track could be followed correctly in projection and its entire length irradiated at the same time. This approach is an alternative to the basic tracking technique but will also entail elaboration if variations in cross section and dose rate along the track length are to be taken care of (see also Section 10).

One further facet warrants mention here—the use of moving couches for irradiating long fields to avoid the problems of joining a series of short fields. Such techniques have been described (Bohndorf and Harder[32]) and may be regarded as a restricted form of tracking.

7. CONSTRUCTING AND INSTALLING THE MARK II TRACKING MACHINE, 1965–1970

A number of events relating to the Tracking Cobalt Project influenced its progress during the five years following its presentation at the Rome Congress in 1965 (Section 5.4). While the machine was under construction at TEM Instruments in Crawley, both its intended location and its prime movers were subject to change.

As a consequence of reorganizing radiotherapy services of the National Health Service in North London, where some centers were to be combined with others, the radiotherapy department at the Royal Northern Hospital was to be phased out and patients treated elsewhere. However, in view of the unique nature of the tracking cobalt machine then under construction, it was agreed that it should be installed in a nearby center, the Royal Free Hospital

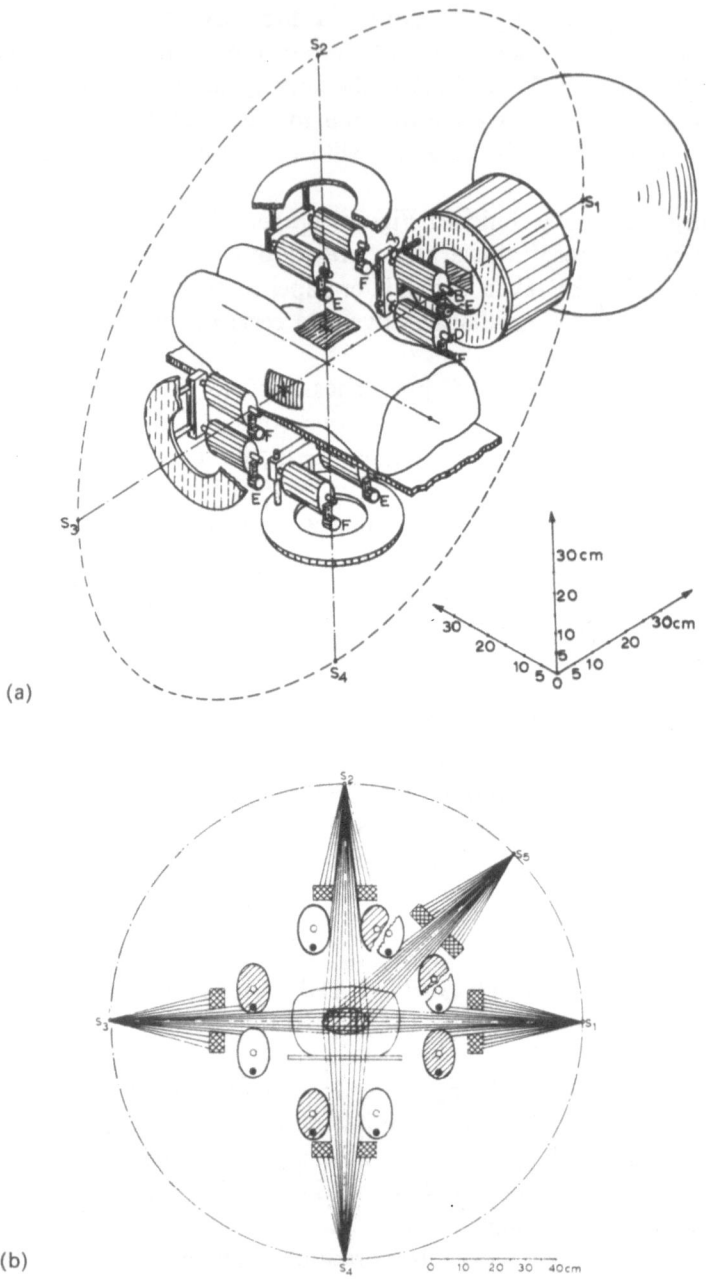

(a)

(b)

Figure 20. Synchronous beam shaping during rotation therapy (Proimos, 1963). (a) Two cylindrical absorbers, oriented continuously by gravity, are shown in the four vertical and horizontal beam positions. (b) Cross section by the vertical $S_1S_2S_3S_4$. The beam width is continuously changed by the gravity-oriented cylinders to fit the width of the tumor.

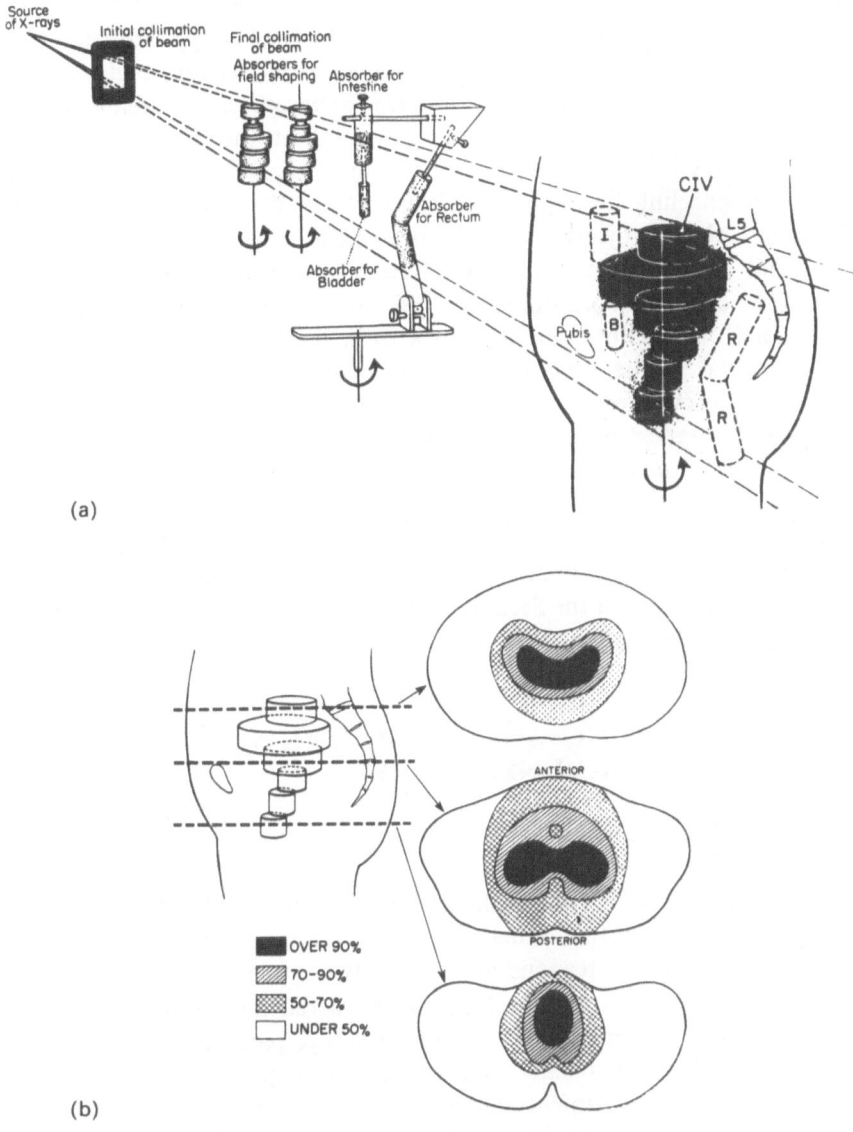

(a)

(b)

OVER 90%
70-90%
50-70%
UNDER 50%

Figure 21. Synchronous shielding and field shaping for rotation therapy of advanced cervical carcinoma (Ilfeld *et al.*, 1971). (a) Oblique view of patient and equipment. The horizontal beam is modified by the field-shaping absorbers to produce the continuously irradiated volume CIV. Absorbers for synchronous shielding protect portions of the small intestine *I*, the bladder *B*, and the rectum *R*. (b) Dose distribution for three pelvic cross sections. The potential tumor volume is within the 90% zone. The size and shape varies at different levels according to the clinical estimate of the tumor volume.

in Hampstead (North London), where a suitably shielded room would be built. Indeed, the machine was duly completed and installed there in 1970.

Regarding individuals, the author (W.A.J.) left his appointment at the Royal Northern Hospital in 1967 to take over responsibility for the Radiation Dosimetry Branch, and later (1976) became superintendent of the Division of Radiation Science and Acoustics at the National Physical Laboratory, Teddington (Southwest London). However, he retained a direct interest in, and link (initially as chairman of an advisory committee to the Tracking Cobalt Project) with, the work of the Royal Free Hospital. T. J. Davy took over responsibility for the physics aspects of the project and thus has led the dosimetry program since 1967. A. Green retired from the Royal Northern Hospital but continued for a while as a consultant at the Royal Free Hospital in the treatment of some patients on the new (Mark II) machine. However, responsibility for the new installation fell to D. B. L. Skeggs, head of the radiotherapy department at the Royal Free Hospital, and he became an enthusiastic supporter of the project and a primary figure in the later development of the Mark III machine (see Sections 8 and 9).

The final Green–Jennings direct presentation relating to the project was entitled the Tracking Cobalt Method or Programmed Three-Dimensional Irradiation, delivered at the Second Congress of the European Association of Radiology, Amsterdam, in 1971 (Green and Jennings[33]).

8. COMMISSIONING AND USING THE MARK II TRACKING MACHINE, 1970–1975

8.1. The Tracking Cobalt Unit

As introduced in Section 5, the tracking cobalt unit (TCU) is a specially modified Mobaltron 100, fitted with servo systems that enable various tracking parameters to respond automatically to program plates, or patient charts, inserted in the control desk. There are two such plates, one example of which is shown in Figure 22, each controlling five parameters, and they are driven forward under a bank of photocells in step with the longitudinal movement of the treatment couch. The program plates are made of translucent Perspex and illuminated from below. The value of each parameter is represented in analogue form as a black profile mounted along the appropriate area on the plate. Above each profile, a photocell is mounted on a point arm, and the deflection of the arm is servo controlled to keep the cell vertically above the edge of the black profile as the plate moves forward. The deflection of the arm operates a control transmitter, a potentiometer, or, in the case of the shutter control, a switch, and these input devices are used to control the machine.

Figure 22. One of two patient charts, containing five control parameters, inserted into the control desk. As the chart progresses into the console, the photoelectric cells follow the "black on white" profiles and operate servo motors to control the treatment machine (see also Figure 15).

Thus, the unit is provided with ten control channels: the *distribution of dose* within a cross section is controlled by the arcing configuration, using five separate channels: the outer sector angle, the outer sector arcing speed, the oscillation axis position, the outer sector beam width, and the inner sector beam width. The *position* of distribution in the cross section is controlled by two channels: the vertical and the lateral displacements. The *dose* is controlled by two channels also: the tracking speed and the beam length. The tenth channel controls the opening of the shutter and the treatment timer.

8.2. Commissioning the TCU

8.2.1. INITIAL INVESTIGATIONS

The machine was installed at the Royal Free Hospital at the end of 1970 and commissioned for "conventional" radiotherapy treatments by May 1971, since the machine can readily be used in that way. Most treatments were delivered on a three-times-per-week basis, making the machine available for tracking investigations on the other two days. The initial investigations comprised: (1) calibrating the various control channels; (2) devising suitable operating procedures; (3) measuring isodose distributions produced by a variety of programs; and (4) assessing the reliability and convenience of the machine.

The program was undertaken by T. J. Davy and his colleagures, and by 1972 a report was prepared for the Tracking Cobalt Project Advisory Committee, recommending certain modifications to the machine. With the help of TEM Instruments, these improvements were put into effect with a view to commencing patient treatments in selected sites.

8.2.2. TREATMENT PLANNING

Localization, entailing AP and lateral radiographs, is necessarily more complex than in conventional treatments because a series of views is necessary along a track to avoid distortion due to beam divergence. A series of patient contours, with reference lines to take care of relative displacements, are needed, and related to setting-up procedures.

Initially, each section is considered as a separate dose distribution within its contour. A number of publications then available provided useful data regarding the optimal placement of arcing axes for cobalt-60 irradiation (e.g., Turner et al.,[34] Tsien et al.[35]). However, the enhanced scope provided by the additional control parameters necessitated a thorough study by both calculation and measurement if the full advantages of TCU were to be realized. This approach is discussed in a paper entitled Conformation therapy using the tracking cobalt unit by T. J. Davy, P. H. Johnson, R. Redford, and J. R. Williams,[11] published in 1975.

In the same paper, Davy et al. discuss achieving a prescribed dose along the track. This depends on the combination of beam length and tracking speed, with a short beam length required if there is much variation in the tumor geometry along the track. Then, to calculate the dose at a point on the track, it is necessary to integrate the cross plot of the beam over the length of the track. The paper[11] should be consulted for a discussion of this computation.

Having decided on the parameters at each contour level, it is necessary to decide how to change them between successive levels, generally by steps, though the positional changes can be made continuously. The photo-cell deflections required to produce a particular value of a parameter can be found from previously measured calibration graphs and program plates prepared.

8.2.3. PROGRAM VERIFICATION

The dose distributions produced by each tracking program are checked by measurement using the Alderson phantom,[19] which provides for sectional distributions at 2.5-cm intervals. This can be achieved using Ilford N4E50 line film, which was found to have a response sufficiently linear up to a dose of 1 gray (100 rad) for cobalt-60 gamma rays. For doses in excess of this amount, brass attenuators can be used to reduce the dose without significantly distorting the distribution. Each film is individually calibrated in terms of dose by using TLD microrods (4% LiF in Teflon). Differences in shape and size of the patient and the phantom are found to have little effect on the dose distribution, but the dose itself will require correction. This can be done by applying calculated tissue–air ratios for the respective cross sections.

8.3. Examples of Treatment Applications

8.3.1. PARTIAL OR COMPLETE USE OF TRACKING FACILITIES

It will be appreciated that the original motivation for developing the tracking technique was the irradiation of lines of spread along sinuous pathways, which requires using all the tracking control parameters. However, the versatility of the Mark II machine lends itself to restricted use by conventional multifield or simple arc treatments as well as to various degrees of elaboration. Thus, to treat a short target volume of elliptical cross section, introducing variable beam width and arcing speed facilities alone may suffice. On the other hand, the longitudinal couch movement alone will be appropriate in some cases. An example of the latter was the treatment of a patient with acute lymphoblastic leukemia, for which the TCU was employed to treat a 42-cm length of spine, using lead blocks at the end of the field to define the end of the treated zone. This is possible when the arcing facility is not required.

In general, when the volume to be treated is irregular, or at a variable depth from the surface, the use of most of all the control parameters will be needed to optimize the distribution. For example, in treating a patient with Hodgkin's disease, a complete tracking technique was used to treat a 25-cm-long volume covering the lymph nodes in the mediastinum, with a sharp cut-off in dose at the superior end of the track but tapered off at the inferior end.

In addition to the original aim of treating sinuous lines of spread, the Mark II machine now encompasses the broader objective of "conformation" therapy. The following examples, taken from the paper by Davy et al.,[11] illustrate the degree of conformation that can be achieved by the TCU.

8.3.2. CARCINOMA OF THE ESOPHAGUS (LOWER HALF)

This is an example of the type of tumor that is well suited to the tracking technique. The tumor is curved in the sagittal plane, and the volume has a fairly small cross section. Multifield treatments or conventional arcing

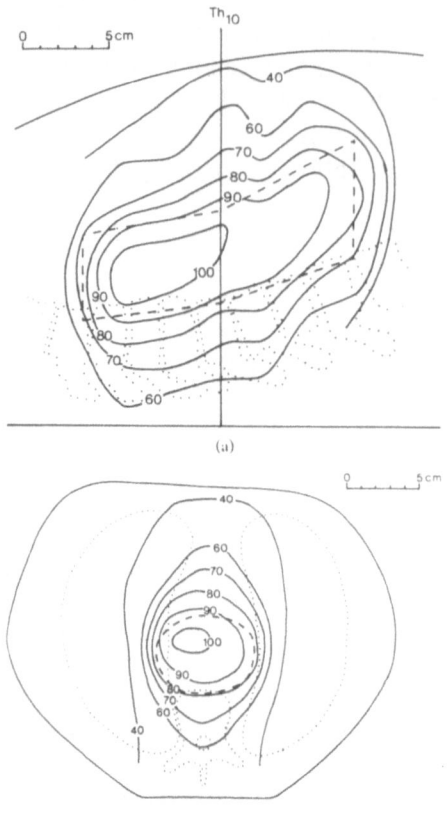

(a)

(b)

Figure 23. A tracking-dose distribution for treating the esophagus (Davy *et al.*, 1975). The dotted line shows the tumor volume. (a) Lateral distribution. (b) Transverse section at midpoint of the tract at level Th 10.

methods would have caused an unnecessarily large volume of healthy tissue to be irradiated at a high dose in order to cover the whole tumor volume. Figure 23(a) shows a lateral view of the isodose distribution, and Figure 23(b) shows the distribution in the central transverse plane achieved by the TCU.

8.3.3. CARCINOMA OF THE RECTUM

The unconventional technique used in this case was chosen to achieve good palliation in spite of the patient's severe ulceration colitis. The tumor was a cylindrical volume, but its axis was inclined at approximately 45° to both the vertical and horizontal planes. If conventional arcing methods had been used, the surrounding bowel would have been irradiated at unacceptably high dose. A tracking technique was therefore prescribed, and AP and lateral views of the isodose distributions are shown in Figure 24(a) and 24(b).

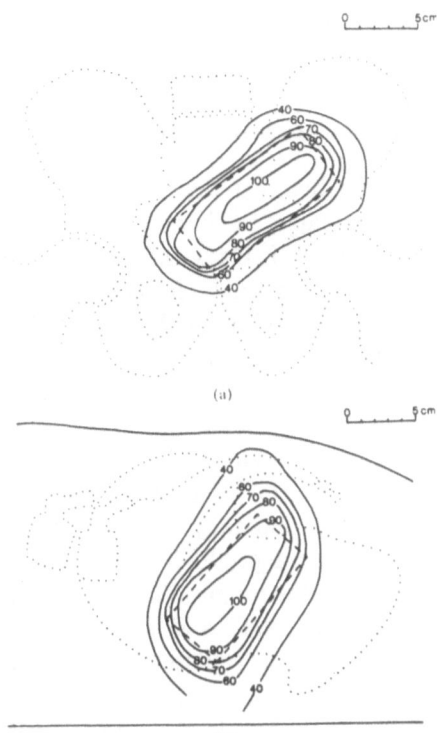

Figure 24. A tracking-dose distribution for treating the rectum (Davy *et al.*, 1975). The dotted line shows the tumor volume. (a) Anteroposterior distribution. (b) Lateral projection.

9. DEVELOPING AND COMMISSIONING THE MARK III TRACKING COBALT MACHINE, 1975–1980

9.1. Limitations of the Mark II Machine

Some 50 patients were treated on the Mark II machine. However, though successful in achieving its objectives, the TCU, with its analogue control system, proved fairly demanding in practice. The advent of the computer provided the opportunity to design a computer-controlled system that would be easier to use and also much more versatile. Moreover, many additional safety features could be introduced. In the TCU, the treatment program was stored in the form of profiles on the two plates in the control unit, and they advanced in synchrony with the treatment couch; in consequence, the couch had to progress in order to change the value of certain parameters. In the computer-controlled approach, the same machine parameters can be used to control the dose distribution, but the values can be changed with time, thereby conferring greater flexibility on the control system. For example, this permits using the machine under computer control for both tracking and conventional treatments. When the machine is set up for a multifield treatment, the patient has to be set up only once, because the machine will proceed automatically to treat a succession of fixed fields.

9.2. Developing the Mark III Tracking Machine

The team at the Royal Free Hospital, comprising D. B. L. Skeggs (radiotherapist in charge), H. S. Williams (head physicist at the Royal Free Hospital), T. J. Davy (physicist in charge of the Tracking Cobalt Project), were joined by J. A. Brace, a computer expert, to develop the computer-controlled tracking cobalt unit (CCTCU) in collaboration with TEM Instruments, which had built the Mark II machine (TCU). TEM Instruments designed the Mark III machine according to the Royal Free Hospital's requirements, which were based on experience gained with the analogue-controlled unit (Mark II).

The CCTCU is a standard TEM MS 90 unit, modified so that it can be used either manually or under computer control. In addition, drives for lateral and longitudinal couch movements were added, and both rate and position of the longitudinal motion can be controlled. A computer interface was incorporated, and some ten parameters were put under computer control: shutter, collimator X, collimator Y, collimator rotation, gantry rotation, head rotation, couch height, couch lateral movement, couch longitudinal movement, and floor rotation. The gantry rotation and couch longitudinal movements are controlled at either a set speed or a set position. The computer also reads the values of these ten parameters, the number of filters inserted, and the angle of rotation of the couch top around the suporting arm.

The CCTCU is controlled by a series of instruction groups or control vectors, each of which describes the next desired position, speed, and shutter state. A computer program reads each control vector from a disk file and translates it into a series of "words" that are then sent to the cobalt unit. Having issued a "move" instruction, the computer monitors the position and state of the cobalt unit every tenth of a second, thus checking when a set of instructions has been obeyed and also detecting any malfunctions of drive that necessitate terminating the treatment. The initial computer system adopted was a Hewlett-Packard System 1000 Model 30 minicomputer with peripherals, used together with a suite of user programs and a set of operating and test procedures.

9.3. Treatment Planning by Computer

While the CCTCU was under development, parallel work was under way at the Royal Free Hospital to introduce computer-aided treatment planning, since here, too, existing procedures for three-dimensional techniques were proving very time consuming. Van de Geijn[20,21,36] had described programs for three-dimensional isodose computations, and cooperation was established with the Department of Clinical Physics and Bioengineering in Glasgow to pursue this approach. Initially, this system proved very helpful, but it was later superseded by alternative developments.

9.4. Funding, Constructing, Installing, and Commissioning the CCTCU

In order to proceed with the Mark III tracking machine, the necessary funding was sought from, and provided by, the Department of Health and Social Security, the Imperial Cancer Research Fund, and the Cancer Research Campaign. A Rad-8 planning system was generously lent by International General Electric (and previously by EMI Medical). A new shielded room was built at the Royal Free Hospital, the machine was constructed at TEM, installed in 1979, and commissioned for treatment in 1980.

This new facility and its use were described by Davy and Brace[37] in a paper presented at the Annual Congress of the British Institute of Radiology in 1979 and in two papers that Brace[38] and Davy[39] presented to the International Symposium on Fundamentals in Technical Progress, Liege, also in 1979.

After commissioning, a progress report on the tracking cobalt project was published in 1981 by the team from the Royal Free Hospital, Brace *et al.*[12]; this described the complete computerized treatment system, treatment planning procedure for the tracking technique, treatment procedure, and some clinical indications of conformation therapy. The computer facet of the technique was also described in another paper by Brace *et al.*[40]

9.5. Clinical Indications of Conformation Therapy

In the preceding progress report (Brace *et al.*[12]), Skeggs stresses that in addition to its superiority over conventional fixed field techniques in treating tumors that are irregular in shape, awkwardly oriented, or tortuous, the CCTCU is also able to treat long volumes at the normal SSD, e.g., medulloblastoma, since the length that can be treated is limited by only the extent of the longitudinal couch movement.

Sites that are especially suitable for treatments by the CCTCU are:

1. Tumors in the neck, such as carcinoma of the thyroid or carcinoma of the upper third of the esophagus, where the disease plunges down in the mediastinum at a very sharp angle.
2. Gynecological tumors extending upward from the pelvis to involve

the para-aortic chain, which can, using this sytem, be treated in continuity with a significantly increased dose.

3. Medulloblastoma and other such tumors, which necessitate irradiation of the hind brain and the whole of the spinal axis.

In short, tumors at any site that run a tortuous course and can now be specifically demarcated with the aid of a computer-assisted tomographic scan are ideally suited for treatment with this system with a greatly improved prospect of success.

9.6. Examples of Dose Distributions Achieved with the CCTCU

The following three sets of isodoses, taken from a paper by T. J. Davy,[39] demonstrate the ability of the new machine to achieve conformation therapy

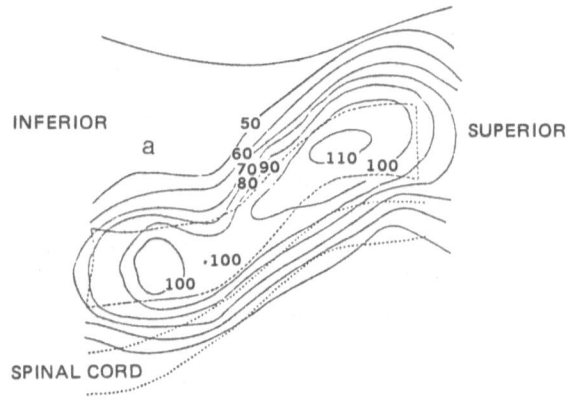

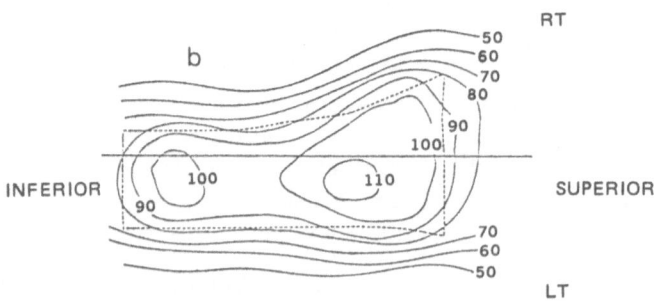

Figure 25. A tracking-dose distribution for treating the esophagus (Davy, 1979). The dashed line shows the tumor volume. (a) Lateral distribution. (b) Anteroposterior projection.

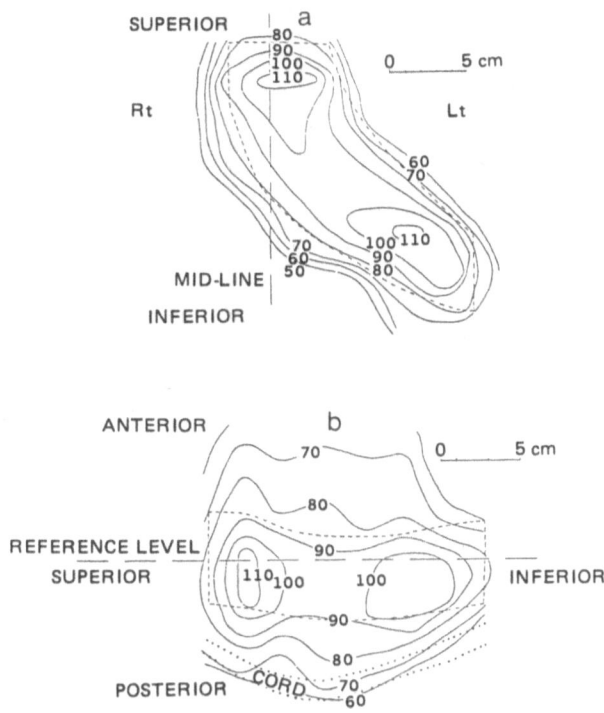

Figure 26. A tracking-dose distribution for treating the bronchus (Davy, 1979). The dashed line shows the tumor volume. (a) Anteroposterior distribution. (b) Lateral projection.

in three dimensions. These examples are related to treatments of the esophagus [Figure 25(a) and 25(b)], the bronchus [Figure 26(a) and 26(b)], and bilateral pelvic nodes [Figure 27(a) and 27(b)]. In each instance, the dashed lines indicate the tumor volume.

9.7. Recent Developments

Problems in three-dimensional dose computations continue to engage the attention of Davy, who presented a discussion on the subject to the Fifteenth International Congress of Radiology in Brussels in 1981.[41] Two basic techniques are considered—one where an arcing technique is used and the other where a number of linear tracks are performed with the source stationary. The use of various control parameters in controlling the shape of the high-dose volume, the dose distribution along the tumor axis, and at the ends of the track, are considered in detail.

The techniques described are equally applicable to linear accelerators,

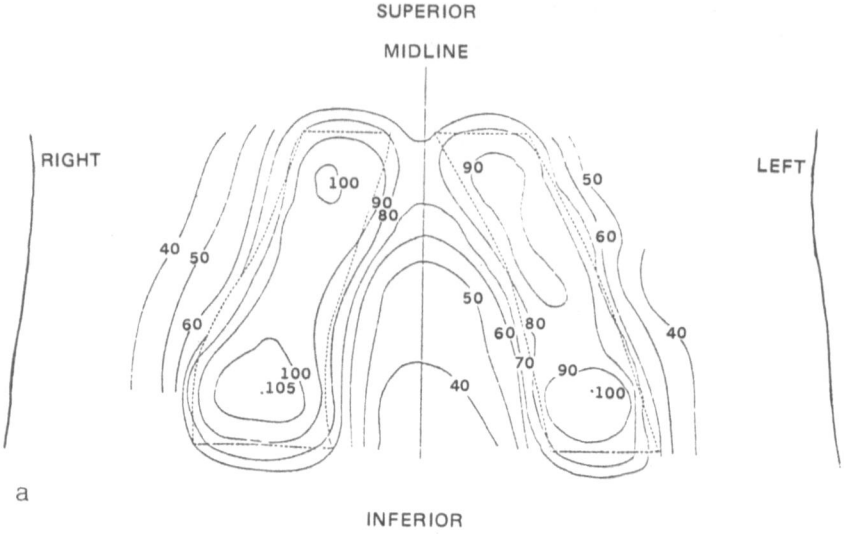

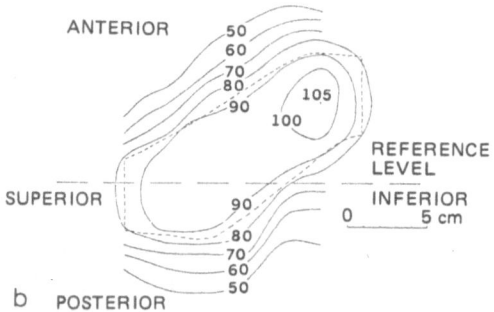

Figure 27. A tracking-dose distribution for treating bilateral pelvic masses (Davy, 1979). The dashed line shows the tumor volume. (a) Anteroposterior distribution. (b) Lateral projection.

and the higher outputs and smaller source size should result in shorter treatment times and a higher degree of conformity.

On the clinical side, Skeggs[42] also presented a paper to the Fifteenth International Congress of Radiology in Brussels in 1981, describing how the Mark III machine could lead to reductions of up to 70% in irradiated volumes for tumors with tortuous shapes, hence making higher doses possible (e.g., esophagus, advanced pelvic tumors). In addition, the crucial importance of computer assisted tomographic scanning in treatment planning was stressed.

The present discussion on the evolution of the tracking cobalt project since its conception concludes at this point, leaving it to Davy and Brace to discuss the present state of the art with respect to treatment-planning procedures and machine control systems in Parts II and III, respectively, of this chapter. However, before closing, a brief reference to parallel work toward conformation therapy elsewhere in the late 1970s is appropriate.

10. ALTERNATIVE APPROACHES TO COMPUTER-CONTROLLED RADIOTHERAPY

In 1978, Kijewski, Chin, and Bjarngard,[43] of the Harvard Medical School, Boston, described the installation of a linear accelerator, modified to allow computer control of several parameters during irradiation of a patient. The initial feasibility study was related to producing wedge-shaped dose distributions by moving collimator jaws. This was followed by a computer-controlled simultaneous variation of a number of treatment parameters, using a Mevatron XII linear accelerator, redesigned for automatic control, and a PDP 11/45 microcomputer, as discussed by Levene et al.[44] The collimator has been modified to allow independent motion of each of the four collimator jaws, with the outer pair being able to move across the center line to permit asymmetric fields. Various techniques have been devised entailing programs where independent variables include the four collimator jaws, the gantry angle, the dose rate, and the couch movements.

Parallel work has been under development in Nagoya, Japan, since it was first introduced there in 1961 by Takahashi.[30,31] A recent paper (1983) by Takahashi et al.[45] lists some 16 references to their work on conformation therapy over the years employing independent collimator jaws, each subdivided into individual sections (up to six) along the longitudinal axis of the field. Thus, the field shape at each gantry angle can be made to match the tumor shape, and it may be asymmetrical with respect to the central axis. The whole system is controlled by individually fabricated cams, and computer programs are introduced for dose computations.

Another center developing dynamic therapy is the Rush-Presbyterian Medical Center, Chicago, where A. Chung Bin has employed a Varian Clinac-4, modified for computer control by using a PDP 11/34 for this purpose. In 1979, Proimos [46] described his "shell technic," a new concept in rotational radiotherapy.

Representatives from these centers in Boston, Chicago and London and others from Japan interested in computer-controlled radiotherapy now hold regular workshops, the first meeting having been in Boston in August 1981 and the second in London in September 1982. Such meetings should expedite progress in overcoming treatment-planning problems and in implementing procedures in commercial systems and thus make conformation therapy practicable in general radiotherapy practice.

ACKNOWLEDGMENTS

The contributions made by a number of individuals to the work discussed in Part I of this chapter have been acknowledged in the text as they arose in the evolution of the project as described. This includes reference to the team engaged in furthering the project at the Royal Free Hospital since the late 1960s.

However, the work of Jennings and Green during the span of over 20 years at the Royal Northern Hospital owed much to the continuing support of the staff in the radiotherapy department.

Assistant physicists, in succession, comprised Ms. B. Burgess, Mr. G. M. Owen, Mr. B. E. Godfrey, Ms. C. Walters, Mr. J. M. Wilkinson, and Ms. R. Lager. On the medical side, assistants included Dr. R. E. Hendtlass, Dr. R. M. Clark, and Dr. J. Stewart; and mention must be made of the radiographers led by Ms. E. V. Webb. In the workshop, Mr. E. V. Marshall, Mr. F. Baker, and Mr. H. A. Holland constructed much ancillary equipment, including the early phantoms. In the experimental program, Mr. H. M. Christie carried out many measurements for the tracking project (Sections 2.3 and 3.3, Ref. 8), and Dr. F. Bush was a consultant physics adviser.

REFERENCES

1. L. J. Shukovsky, Dose, time, volume relationships in squamous cell carcinoma of the supraglotic-larynx, *Am. J. Roentgenol.* **108**, 27–29 (1970).
2. J. G. Stewart and A. W. Jackson, The steepness of the dose-response curve both for tumor cure and normal tissue injury, *Laryngoscope* **85**, 1107–1111 (1975).
3. D. F. Herring, The consequences of dose-response curves for tumor control and normal tissue injury on the precision necessary in patient management, *Laryngoscope* **85**, 1112–1118 (1975).
4. A. Green, A technical advance in irradiation technique, *Proceedings of the Royal Soc. Med.* **52**, 344–346 (1959).
5. A. Green, The directional caliper: its clinical uses in radiation therapy, *Br. J. Radiol.* **10**, 95–101 (1937).
6. J. L. Dobbie, Beam direction in X-ray therapy, *Br. J. Radiol.* **12**, 121–128 (1939).
7. A. Green, W. A. Jennings, and F. Bush, Rotational Roentgen therapy in the horizontal plane, *Acta Radiol.* **31**, 275–320 (1949).
8. A. Green, W. A. Jennings, and H. M. Christie, Radiotherapy by tracking the spread of disease, *Transactions of the Ninth International Congress of Radiology, Munchen 1959*, Verlag, Stuttgart, 766–772 (1960).
9. W. A. Jennings, Programmed three-dimensional irradiations, *Abstract of the Eleventh*

International Congress of Radiology, Rome 1965, International Congress Series No. 89, Excerpta Medica Foundation, Amsterdam 332 (1965).

10. A. Green, Tracking cobalt project, *Nature* **207**, 1311 (1965).

11. T. J. Davy, P. K. Johnson, R. Redford, and J. R. Williams, Conformation therapy using the tracking cobalt unit, *Br. J. Radiol.* **48**, 122–130 (1975).

12. J. A. Brace, T. J. Davy, D. B. L. Skeggs, and H. S. Williams, Conformation therapy at the Royal Free Hospital. A progress report on the tracking cobalt project, *Br. J. Radiol.* **54**, 1068–1074 (1981).

13. R. Knox and A. St. G. Caulfield, described in *A Textbook of Radiotherapeutics*, 4th ed., W. M. Levitt, ed, p. 79–80, A. & C. Black, London (1932).

14. U. Henschke, Uber Rotationsbestrahlung, *Fortschr. auf dem Gebiete der Röntgenstr.* **58**, 456–461 (1938).

15. H. Meyer, Das Problem der "Kreuzfeuerwirkung" in der gynäkologischen Roentgentherapie, *Zentralbl. Gynäk.* **37**, 1741–1752 (1913).

16. J. Nielsen and S. H. Jensen, Some experimental and clinical lights in the rotational therapy and its basis and possibilities, *Acta Radiol.* **23**, 51–66 (1942).

17. W. A. Jennings and A. L. McCrea, Dose distribution in conical rotation therapy with a 2-MeV generator, *Radiology* **68**, 689–697 (1957).

18. W. A. Jennings, Percentage depth dose in moving-field therapy, *Radiology* **68**, 698–707 (1957).

19. S. W. Alderson, L. H. Lanzl, M. Rollins, and J. Spira, An instrumented phantom system for analog computation of treatment plans, *Am. J. Roentgenol.* **87**, 185–195 (1962).

20. J. van de Geijn, Dose distribution in moving-beam cobalt-60 teletherapy—a generalised calculation method, *Br. J. Radiol.* **36**, 879–885 (1963).

21. J. van de Geijn, The computation of 2- and 3-dimensional dose distributions in cobalt-60 teletherapy, *Br. J. Radiol.* **38**, 369–377 (1965).

22. K. A. Wright, B. S. Proimos, J. G. Trump, M. I. Smedal, D. O. Johnson, and F. A. Salzman, Field shaping and selective protection in megavolt radiation therapy, *Radiology* **72**, 101 (1959).

23. B. S. Proimos, Synchronous field shaping in rotational megavolt therapy, *Radiology* **74**, 753–757 (1960).

24. B. S. Proimos, Synchronous protection and field shaping in cyclotherapy, *Radiology* **77**, 591–599 (1961).

25. B. S. Proimos, New accessories for precise teletherapy with cobalt-60 units, *Radiology* **81**, 307–316 (1963).

26. B. S. Proimos, Beam shapers oriented by gravity in rotational therapy, *Radiology* **87**, 928–933 (1966).

27. B. S. Proimos, S. P. Tsialas, and S. C. Coutroubas, Gravity-oriented filters in arc cobalt therapy, *Radiology* **87**, 933–937 (1966).

28. B. S. Proimos, Shaping the dose distribution through a tumor model, *Radiology* **92**, 130–135 (1969).

29. D. N. Ilfeld, K. A. Wright, and F. A. Salzman, Synchronous shielding and field shaping for megavolt irradiation of advanced cervical carcinoma, *Am. J. Roentgenol.* **112**, 792–796 (1971).

30. S. Takahashi, T. Kitabatake, K. Morita, S. Okajima, and H. Iida, Methoden zur besseren Anpassung der Dosisverteilung an tiefliegende Krankheitsherde bei Bewegungsbestrahlung, *Strahlentherapie* **115**, 478–488 (1961).

31. S. Takahashi, Conformation radiotherapy: rotation techniques as applied to radiography and radiotherapy of cancer, *Acta Radiol. Suppl.* **242**, 1–142 (1965).

32. W. Bohndorf and D. Harder, Die Dosisverteilung bei der horizontaltranslation an Telekolaltgeraten, *Strahlentherapie* **119**, 389–400 (1962).

33. W. A. Jennings and A. Green, The tracking cobalt-60 method, or programmed 3D irradiation, *Abstracts of the Second Congress of the European Association of Radiology, 1971*, Excerpta Medica International Congress Series, no. 230, p. 154, Excerpta Medica, Amsterdam (1971).

34. J. E. Turner, R. M. Johnson, and S. M. Whitfield, An analysis of factors affecting optimal axis placement and 80% isodose volume dimensions in telecobalt arc therapy, *Am. J. Roentgenol.* **94**, 852–864 (1965).

35. K. C. Tsien, J. R. Cunningham, and D. J. Wright, Effects of different parameters on dose distributions in cobalt-60 planar rotation, *Acta Radiol. Ther. Phys. Biol.* **4**, 129–154 (1966).

36. J. van de Geijn, A computer programme for 3-D planning in external-beam radiation therapy, EXTDOS, *Comput. Programs Biomed.* **1**, 47–57 (1970).

37. T. J. Davy and J. A. Brace, Dynamic 3-D treatment using a computer-controlled cobalt unit, *Br. J. Radiol.* **53**, 384 (1979).

38. J. A. Brace, A computer system for the dynamic control of a telecobalt unit, *International Symposium on Fundamentals in Technical Progress, Liege, Belgium*, Presses Universitaires de Liege (1979).

39. T. J. Davy, Dynamic treatment using a computer-controlled telecobalt-60 unit, *International symposium on fundamentals in technical progress, Liege, Belgium*, Presses Universitaires de Liege (1979).

40. J. A. Brace, T. J. Davy, and D. B. L. Skeggs, Computer-controlled cobalt unit for radiotherapy, *Med. Biol. Eng. Computing* **19**, 612–616 (1981).

41. T. J. Davy, The control of radiotherapy dose distributions in three dimensions using a computer-controlled tracking unit, *Abstract, Int. Congress Radiol., Brussels* (1981).

42. D. B. L. Skeggs, 3-dimensional radiotherapy under computer control and its relationship to CT scanning, *Abstract, Int. Congress Radiol., Brussels* (1981).

43. P. K. Kijewski, L. M. Chin, and B. E. Bjarngard, Wedge-shaped dose distributions by computer-controlled collimator motion, *Med. Phys.* **5**, 426–429 (1978).

44. M. B. Levene, P. K. Kijewski, L. M. Chin, B. E. Bjarngard, and S. Hellman, Computer-controlled radiation therapy, *Radiology* **129**, 769–775 (1978).

45. K. Takahashi, J. A. Purdy, and Yeong Ylin, Work in progress: treatment planning for conformation radiotherapy, *Radiology* **147**, 567–573 (1983).

46. B. S. Proimos, "Shell technic," a new concept in rotational radiotherapy, *Proceedings of the Fifth International Conference on Medical Physics, and the Twelfth International Conference on Medical and Biological Engineering, Jerusalem*, Bellinson Medical Center, Petak, Tikra, Israel (1979).

1-II

Physical Aspects of Conformation Therapy Using Computer-Controlled Tracking Units

T. J. DAVY

1. INTRODUCTION

As discussed by W. A. Jennings in Chapter 1-I, the term *conformation therapy* is used to denote treatments in which the high-dose volume is shaped in three dimensions to match the target volume. The term also implies controlling the dose distribution throughout this volume and the surrounding healthy tissue. The immediate technical objective is to minimize the high-dose volume in order to increase the tolerance dose; reduce the dose to sensitive organs, such as the spinal cord, kidneys, and lungs lying adjacent to the target volume; and minimize the integral dose to the patient. Such treatments are carried out using dynamic treatment techniques, a term that is used in this chapter to describe treatments employing complex beam arrangements that are executed by automatic machines in response to some stored treatment file. Such files may be stored in analogue form, as in treatments using synchronous shielding or in digital form in machines controlled by mini- or microcomputers.

T. J. DAVY ● Medical Physics Department, Royal Free Hospital, Pond Street, Hampstead, London NW3 2QG England.

Despite the anticipated clinical advantages, few centers are using conformation therapy at present. There are a number of historical reasons for this, including the difficulties in defining the target volume in three dimensions, the complexity and expense of the equipment, and the complications of three-dimensional treatment planning. Developments in computer technology have introduced the computerized tomography (CT) scanner into the hospital, simplified the automation of treatment machines, and improved treatment-planning systems. The time seems right for a reappraisal of conformation therapy methods and increased efforts to overcome treatment-planning difficulties.

This section describes some of the basic principles of conformation therapy and outlines treatment methods currently in use or being developed at the Royal Free Hospital (RFH). These methods are dictated by the operational modes of the treatment unit, and the treatment method chosen determines the approach to treatment planning and dose calculation.

Conventional treatment-planning systems are, in fact, simply dose-calculation systems that rely heavily on the expertise of the human planner to produce a suitable dose distribution.[1] For single-plane plans using a small number of treatment beams, planning can be carried out interactively, with the planner observing the dose distribution and then adjusting the position of the beams or changing the dose weightings from the various beams. Such methods are unlikely to prove successful for complex three-dimensional treatment plans, partly because of the number of beams and the lengthy computing time but principally because of the difficulty in estimating the effect of changing a beam configuration designed for one plane on the dose distribution in adjacent planes.

In routine planning for conformation therapy at the RFH use is made of exposure time profiles and (absorbed) dose profiles. Although such profiles are only determined for the patient's superior–inferior axis at present, it is believed that these profiles may be usefully linked with transverse plane profiles to describe operations of the treatment unit in three dimensions. For this purpose, the dynamic treatment unit may be considered as a three-dimensional exposure time controller, and the optimization of three-dimensional plans may ultimately be based on machine parameters rather than isodose curves.

The selection of a three-dimensional treatment plan involves considering the best attainable dose distribution and how it can be achieved by the treatment machine. Treatment-planning strategies are therefore discussed in terms of "thin-slice" and "thick-slice" planning for dose distribution purposes and in terms of a "field-by-field" or "slice-by-slice" approach to practical planning and machine operation.

In order to plan a patient's treatment in three dimensions, the patient's geometry must be measured in three dimensions together with physical data, such as electron densities needed for dose calculations. A planning and tumor localization procedure using a CT scanner is described. A vital part of the overall planning process is treatment simulation. Conventional simulators are unable to simulate tracking treatments and at present, it is necessary to rely on measurements being made on phantoms using film, TLD dosimeters, and other probes. An Alderson Rando phantom[2] is used for full three-dimensional studies, but a simpler phantom is used to measure the axial dose profiles and thus test those parameters that control the axial dose profile. For measurements with the axial dose profile phantom, specially modified versions of the actual treatment files are used.

Quality assurance for tracking units and other computer-controlled machines requires understanding the operating principles of the unit. It is important to appreciate that the controlling computer relies on data fed to it from the machine transducers and that routine physics and service tests should verify the accuracy of data from the machine. The simple axial dose profile phantom may be used as a "black box" test of the dose-controlling parameters. A simple unit-density phantom may be used to confirm that the three-dimensional plan being performed by the treatment unit is the same as that produced by the planning system.

2. METHODS FOR ACHIEVING CONFORMATION THERAPY USING PHOTON BEAMS

2.1. Basic Requirements for Controlling Dose Distribution in Three Dimensions

In order to produce an ideal three-dimensional dose distribution in the patient (assuming that it is known), the treatment machine needs certain facilities to produce it, viz.:

1. Perfect control of the shape of the beam cross section.
2. Perfect control of the dose distribution over the beam cross section.
3. Perfect control of the dose distribution in the direction parallel to the central ray.

Such a machine, would, in fact, permit treatment of any tumor with a single beam.

The most significant factor in controlling the shape of the high-dose volume is the shape of the radiation field. In this context, the term *beam* is used to denote the cone of photons emerging from the machine collimators, whereas the term *field* refers to the aggregate effect of the beam or beams at the target after additional shaping or combining by shielding blocks or machine movement.

The effects of beam weighting on the shape and size of the high-dose zone in a single transverse plane are well known and can easily be explored using conventional treatment-planning systems. The incident beam profile may be modified by using wedges or compensators. For conformation therapy, the dose distribution may change from one transverse plane to the next, and the dose weighting must be controlled along the tumor axis. Control over the dose distribution in a direction parallel to the central ray, however, is very limited. Small changes in the depth dose curve may be made by varying the treatment distance or the beam energy, or by adding bolus if an additional dose is required at the surface.

The mantle treatment described by Walbom-Jorgensen et al.[3] illustrates a treatment in which the field shape is well controlled in the coronal plane by a specially cast shielding block and the axial dose distribution is controlled by a compensator. There is, however, little control over the dose distribution in the anterior–posterior direction. In contrast to this, many conventional treatment plans using conventional cross-fire techniques produce satisfactory dose distribution in a single transverse plane for tumors of simple shape. For conformation therapy, it is necessary to combine the features of these two techniques and achieve a greater degree of control over the dose distribution in the transverse plane in order to cope with target cross sections of complex shape.

2.2. A Brief Comparison of Conformation Therapy Systems

The basic requirement for conformation therapy is that the radiation field must match the projected tumor outline for all beam entry angles. This aim may be achieved by three basic methods, viz., by using multisegment collimators, synchronous shielding, or the tracking technique described in the first section. These systems are briefly compared in the following sections.

2.2.1. MULTISEGMENT COLLIMATOR MACHINES

Multisegment collimator machines use collimators with each width-defining blade made up of a number of narrow "fingers," each of which can be moved in or out of the beam emerging from the primary collimator under the control of an analogue or a digital control system.[4–6] Treatment times are similar to those of conventional units. The resolution in terms of geometrical fit between the target and the high-dose volume is limited by the projected width of the individual segments, which is typically 3–4 cm at the isocenter. Treatment planning is straightforward from a geometrical point of view, and there are no increased collision risks between couch and patient. Wedged fields may be produced by collimator adjustments. The system lacks any inherent facility for controlling the axial dose distribution, and the maximum length of target volume is limited by the available beam size as for conventional units. Adjacent collimator segments are interlocked to minimize leakage between them.

2.2.2. SYNCHRONOUS SHIELDING

As described in the first part of this chapter, this system requires mechanical devices attached to the head of the treatment unit and in principle may be used with almost any supervoltage unit if there is sufficient clearance around the couch and the patient.[7-10] The exposure times are the same as for conventional treatments, but time is required to mount the accessories and construct them for each patient. Treatment planning is essentially empiric, since the beam profile is continuously changing during the exposure—the beam shaping and shielding blocks present a differently shaped edge to the beam at each gantry angle. The facility for shielding structures in, or adjacent to, the target volume is an attractive feature of the system, but again there is no intrinsic facility for controlling the dose distribution along the tumor axis.

2.2.3. TRACKING SYSTEMS

Tracking machines are so named because the treatment couch is moved relative to the machine isocenter so that the isocenter follows or tracks along some predetermined locus in the patient. Such tracking units as the TEM MS90 computer-controlled tracking cobalt unit at the RFH[11-13] and the computer-controlled Varian Clinac-4 described by Chung-Bin et al.[14] are based on standard treatment units adapted for computer control. In these machines, the beam is positioned in the patient's cross section by adjusting the couch's vertical and lateral positions. A computer-controlled Mevatron XII linear accelerator (linac) described by Bjarngard and other authors [15-17] is fitted with special collimators where each outer collimator blade is independently controlled, so that the beam can be swept across the source rotation plane. Since these blades can pass over the center line of the collimator assembly, the beam can be used to irradiate targets of a crescent-shaped cross section in the transverse plane using tangential fields. This machine is also used to carry out a dynamic-wedging technique where the collimator is slowly opened, with one blade remaining stationary, to produce the required dose gradient. Crescent-shaped target volumes can be irradiated using the basic tracking units by suitable couch displacements as discussed in Section 7.4; the dynamic wedging technique referred to above, however, is not practical on cobalt units because of the large source size and the attendant minimum usable beam size of 3 cm or so.

Tracking machines possess an inherent facility for controlling the axial dose distribution to compensate for variations in tumor depth and the presence of inhomogeneities. The dose delivered to a point passing through the treatment beam is proportional to the exposure time and so is proportional to the beam length (measured in the direction of axial couch movement) and inversely proportional to the speed of axial couch movement (termed the tracking speed).

Since irradiated fields or volumes are built up by a tracking or scanning process, the size of the tumor that may be irradiated at the isocenter is limited by only the range of couch travel. The closeness of fit between target volume and high-dose zone is determined by the beam length used. The exposure time for each track will be greater than that for an exposure of the same area using a single large beam by the ratio of the length of the target volume to that of the beam length ("scanning slit width"). Since the total exposure time is a significant factor in designing treatment room protection, this factor should be taken into account if tracking treatments are contemplated. From the clinical point of view, any increase in leakage contribution to the patient must be considered and balanced against the considerable reduction in high-dose volume and the associated scatter when conformation therapy is used.

The sections that follow describe the RFH approach to treatment planning and treatment on the MS90 CCTCU that has now been in use under computer control for patient treatment since January 1980, with system software designed and implemented by Brace.[13] The principles and methods described are applicable to cobalt units and linacs, although there are some significant differences between them. The major difference for control purposes is that the cobalt unit has a steady dose rate and may be controlled by a timer, whereas the linac output may vary and is therefore controlled by pulses from the transit dosimeters in the head of the machine. For treatment-planning purposes, the most significant differences are the improved geometrical fit between the target volume and the high-dose volume, due to the shorter usable beam length of the linac. This shorter beam length also improves the resolution of the axial dose distribution. The higher energy of the linac beam and the higher dose rates serve to reduce integral dose and reduce the overall exposure time as in conventional treatments. The penumbra of the cobalt beam simplifies butting together numbers of stationary beams but is not generally advantageous.

3. REPRESENTING THREE-DIMENSIONAL TREATMENT PARAMETERS

3.1. The "Ideal" Beam

The simple-beam model shown in Figure 1 is used to illustrate various aspects of tracking treatments. The "ideal" beam has an effective length X_E

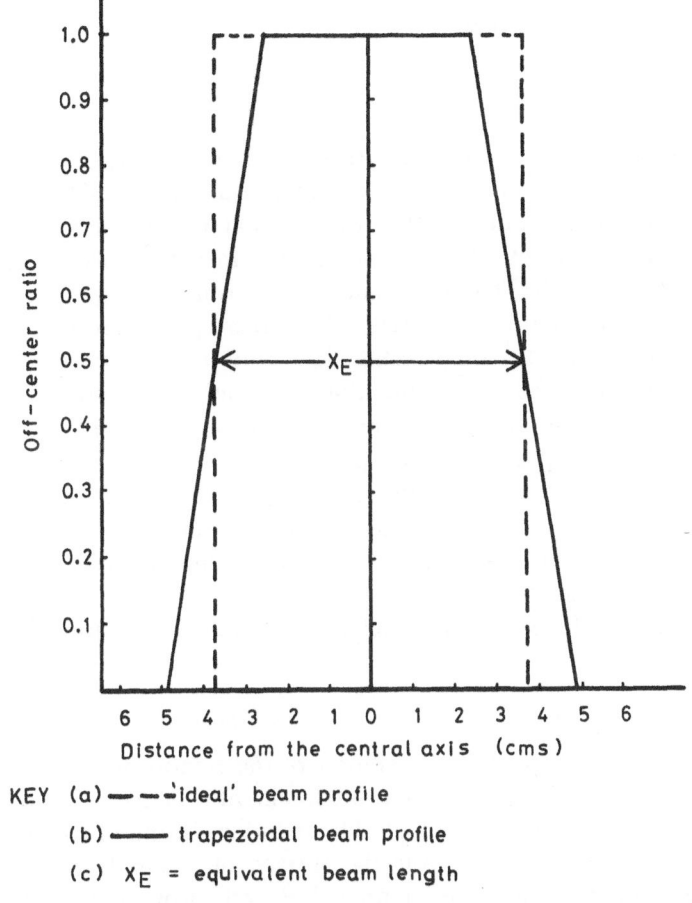

KEY (a) ── ──'ideal' beam profile

     (b) ──── trapezoidal beam profile

     (c) X_E = equivalent beam length

Figure 1. Diagram illustrating the simple models used to represent the dose-rate profile at the isocenter in the principal planes of the treatment beam. The effective beam length X_E is approximately equal to the 50% beam width.

and an off-center ratio[18] of 1.0 for a distance of $X_E/2$ either side of the central axis, dropping immediately to 0.0 outside these limits. The area under the beam profile equals that of the actual beam profile when integrated between specified limits. X_E is approximately equal to the 50% beam size when integrated over a range extending for 10 cm either side of the useful beam (80% definition). The actual value of X_E will depend on the integration limits and other factors, such as the thickness of overlying material as discussed by Williams[19] and Davy et al.[20] Values of X_E for various collimator settings may be obtained by direct measurement with tracking units or may be calculated by integrating under the dose-rate profiles using various beam models such as that developed by van de Geijn.[21]

3.2. Exposure-Time Profiles, Exposure-Dose Profiles, and Absorbed-Dose Profiles

3.2.1. DEFINITIONS

An exposure-time profile (ETP) is a plot of the exposure time (ET) against the position along the same locus. Exposure-time profiles are suitable for describing the operations of therapy units that have a steady dose rate and may be controlled by exposure timers; for machines producing pulsed radiation beams controlled by transmission exposure meters, exposure (dose) profiles (EDP) are more appropriate. While these profiles resemble conventional beam profiles to some extent, they show integrated values along a chosen locus rather than dose rates along a straight line. In this context, the term *exposure* is used in the quantitative sense ($X = dQ/dm$).

The exposure pattern in the patient will result in a corresponding (absorbed) dose pattern that may also be expressed as a series of absorbed-dose profiles (ADPs). In general, the distribution will be calculated along the same loci as the exposure profiles. The general form of the ETPs, EDPs, and the ADPs will be similar, and for simplicity, ETPs will be used to illustrate the use of profiles for expressing machine operations and treatment planning.

3.2.2. AXIAL ETPs

Axial ETPs describe the operation of the treatment unit in terms of exposure times along loci running predominantly in the superior–inferior directions. Figure 2 shows the ETP due to an ideal beam of constant length traveling at uniform speed over the surface of a simple phantom. The corresponding ADP may be computed directly from the primary-beam profile using scatter-air ratios. Alternatively, conventional beam dose rate profiles measured in water phantoms may be used with integration being performed under the beam profile between the desired limits.

3.2.3. TRANSVERSE-PLANE ETPs

Transverse-plane ETPs describe the operation of the treatment unit in terms of exposure times along loci in the patient's cross section. Usually, loci will be selected as closed loops running around the perimeter of the target, but other loci may be chosen to run through regions of interest, such as the kidneys. Figure 3 shows some ETPs for a simple rotation treatment, and Figure 4 shows ETPs for a three-field cross-fire technique.

If lines are drawn around the tumor in a series of transverse planes along the superior–inferior axis of the tumor and the ETs and EDs plotted against the position along the lines (e.g., at 10° intervals), an ET or ED pattern results

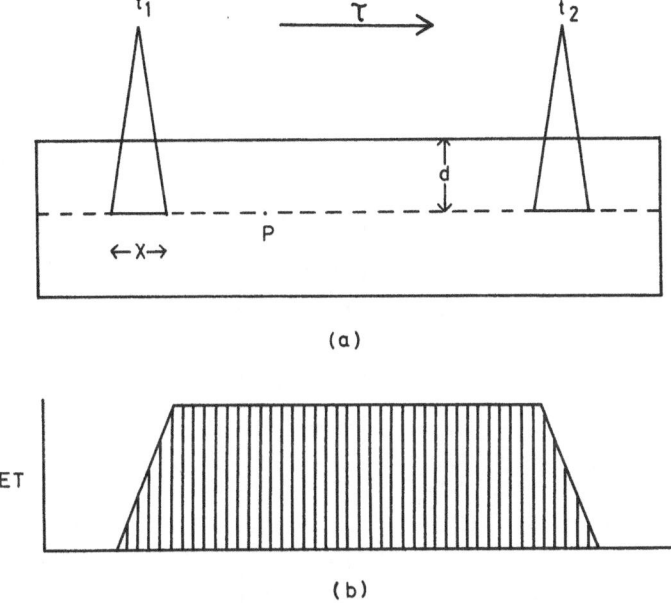

Figure 2. The integrated exposure-time profile produced by a simple linear track. (a) Movement of an "ideal beam" of fixed length X along a phantom at constant speed τ. Movement starts at time t_1 and ends at time t_2. (b) The integrated exposure-time profile along the axis of the phantom at depth d.

that represents beam movements and time or exposure weighting around the target volume. These loops may be linked by a line or lines representing profiles along the tumor axis or running along the surface of the target. The target volume may thus be enclosed in a "cage" of loci for exposure profiles as shown in Figure 5.

The ADPs may be computed along the same loci as the exposure profiles and used to represent the dose distribution. Conventional isodose curves may, however, be preferred to show the final dose distribution. If the required dose distribution is specified for a series of a parallel planes, the dose distribution needed between these planes may be inferred. If a series of single-plane profiles is chosen to represent the best attainable distribution in each plane, then the combined three-dimensional pattern will show the best three-dimensional plan if the limits of resolution of the treatment beam along the superior–inferior axis are disregarded. The three-dimensional optimization of the plan may therefore be based on obtaining the best match between the set of ideal single-plane profiles and those attainable by the treatment unit.

If the transverse-plane ADPs are converted to the corresponding ETPs or EDPs, they may be regarded as the best attainable set of exposure conditions ignoring the limits imposed by the use of a beam of finite length.

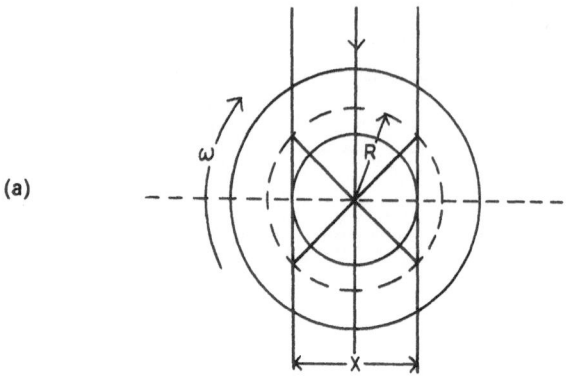

(a)

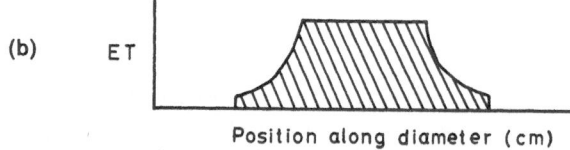

(b)   ET

Position along diameter (cm)

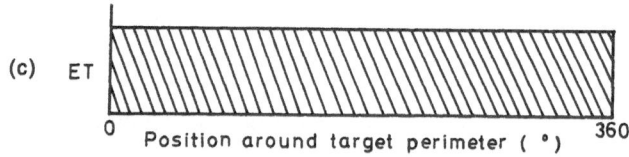

(c)   ET

0   Position around target perimeter (°)   360

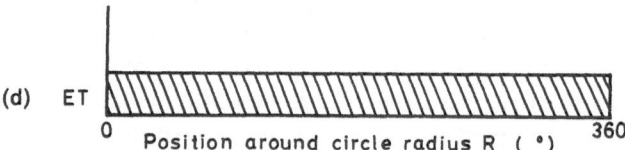

(d)   ET

0   Position around circle radius R (°)   360

Figure 3. Beam configuration and exposure-time profiles for one complete rotation. (a) Single beam irradiating a circular phantom. (b) Exposure-time profile along any diameter. (c) Exposure-time profile around the target perimeter. (d) Exposure-time profile around a circle of radius R ($R > 0.5X$).

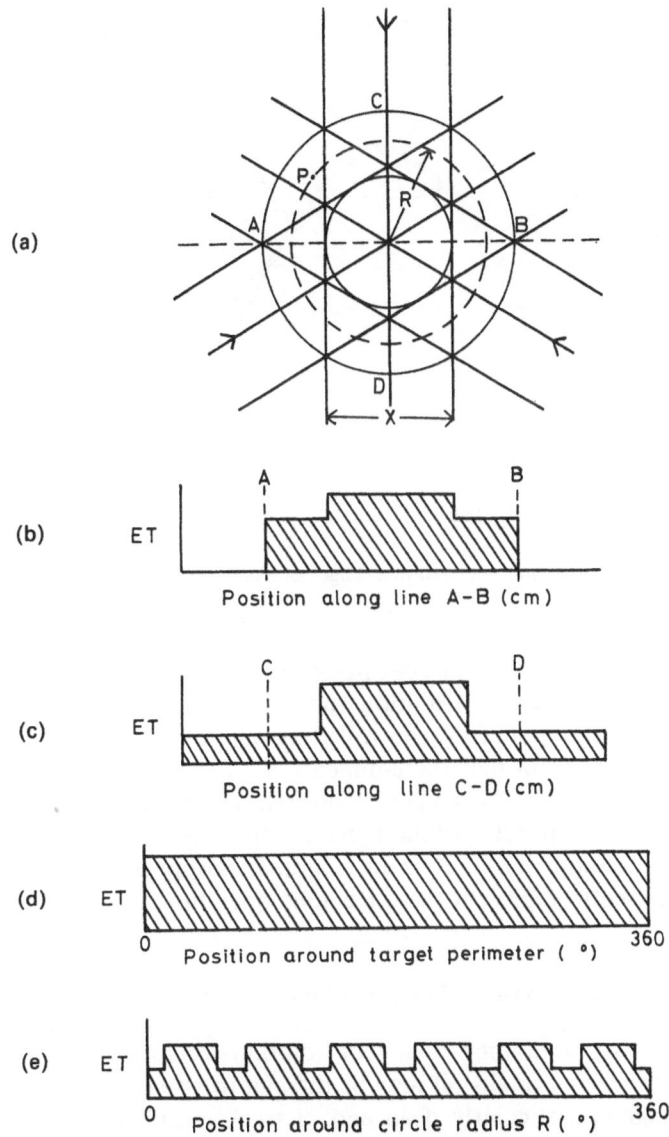

Figure 4. Beam configuration and exposure-time profiles for a symmetrical three-beam irradiation of a circular phantom. (a) Three beams irradiating a circular phantom. (b) Exposure-time profile along AB axis. (c) Exposure-time profile along XY axis. (d) Exposure-time profile around the target perimeter. (e) Exposure-time profile around a circle of radius $R(R > 0.5X$ and $< X)$.

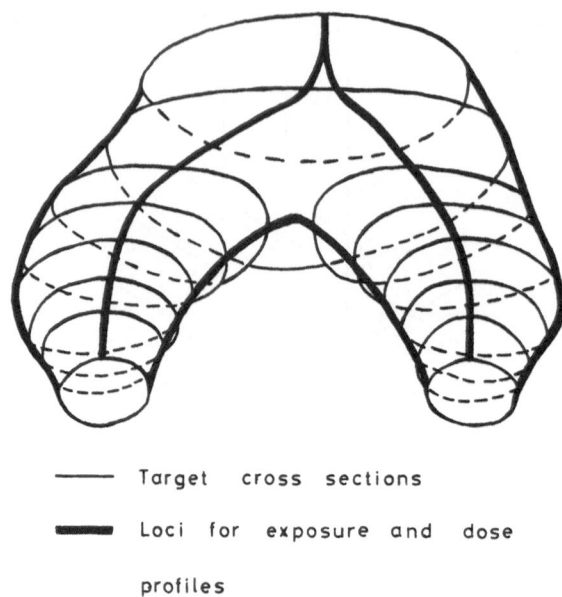

———— Target cross sections

▬▬▬ Loci for exposure and dose

profiles

Figure 5. Complex target volume enclosed in a "cage" of loci for exposure and dose profiles.

The optimization process may thus be directed toward finding the best match between this ideal exposure distribution and that attainable in practice.

Since the exposure pattern is controlled by only the treatment machine parameters, it would seem that three-dimensional optimization procedures may be based on a process designed to produce the best compromise between those machine operations selected for a series of thin slices and an attainable set taking into account the finite-beam length. Exposure-time profiles and EDPs may be used to represent these machine operations.

3.3. Radial Time and Exposure-Weighting Diagrams

For treatments combining arcing or rotation techniques with couch axial translation (arcing tracks), it may be difficult to visualize the effects of various parameters, particularly if an end of track technique is used (see Section 4.4) together with variations in gantry speed and arcing angles.

A machine carrying out a rotation at constant angular speed around a fixed axis in the phantom (or space) while the couch proceeds at a constant speed may be represented in part by a series of radial lines whose length is proportional to the time the beam remained in the sector. This is somewhat similar to selecting "stations" for beams in rotation techniques in conventional planning systems.

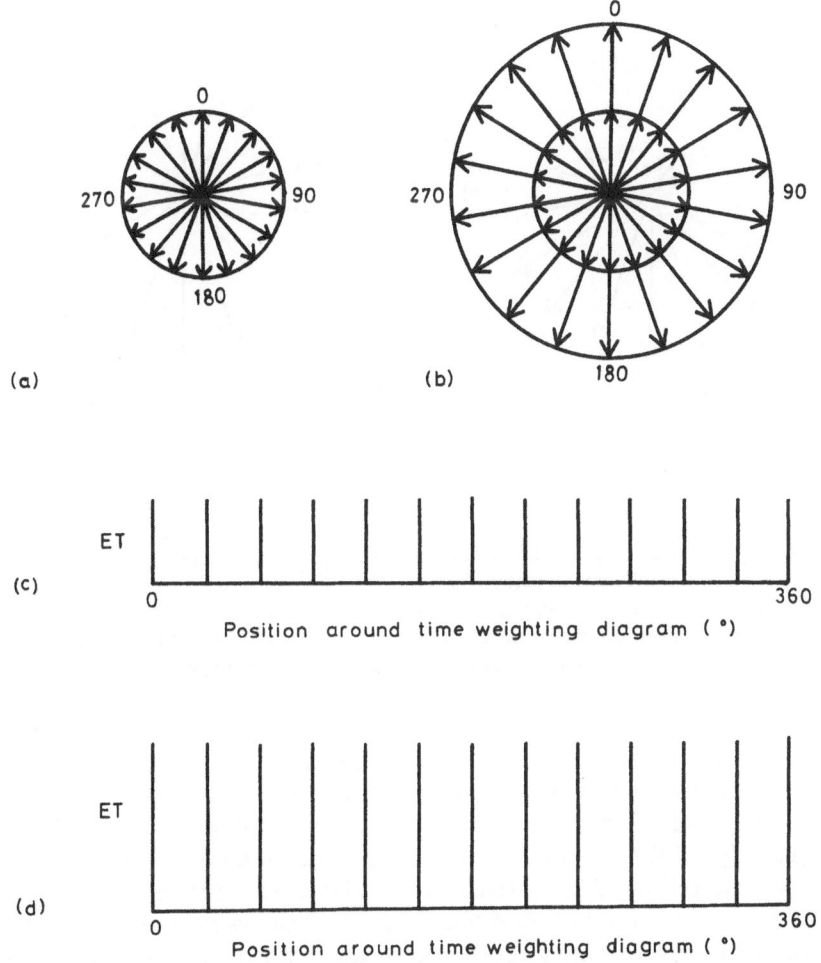

Figure 6. Diagram showing exposure-time weightings for various angles of beam entry. (a) Weightings for one complete rotation at 20° intervals. (b) Weightings for two rotations at 20° intervals. (c) Weightings for one complete rotation displayed at 20° intervals on a linear scale. (d) Weightings for two rotations displayed at 20° intervals on a linear scale. In diagrams (a) and (b) the exposure times appropriate to the various angles are represented by the radial line lengths.

If an ideal beam is considered where the calculation plane passes through the beam during one complete revolution, the pattern shown in Figure 6(a) is obtained when lines are drawn at intervals of 20°. Figure 6(b) shows the representation of two complete rotations. These weightings may also be represented on a linear scale as in Figure 6(c) and 6(d).

In practice, the situation is more complex. The effect on weighting due to the transit of beams of different profiles through the calculation plane is

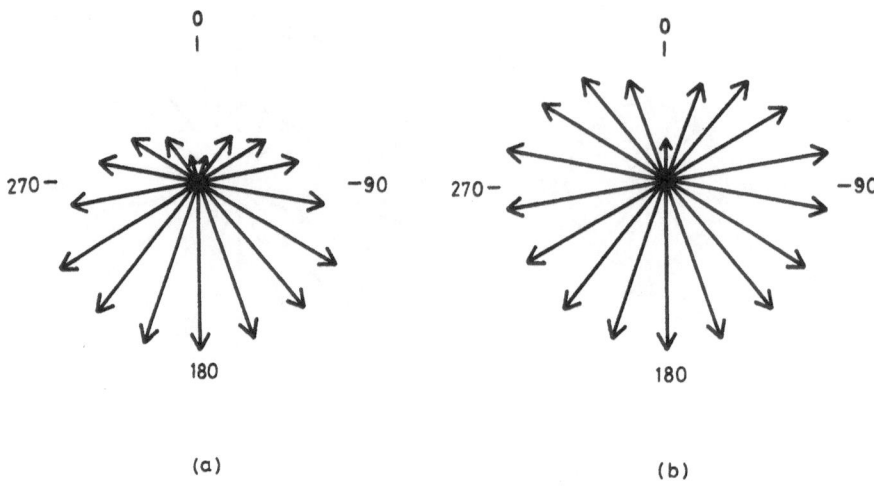

Figure 7. Diagram showing the effect of a trapezoidal-beam profile on the radial exposure-time weighting pattern. (a) Weighting pattern for one complete rotation of the gantry during passage of a beam through the source rotation plane. (b) Weighting pattern for one complete rotation of the gantry during passage of a trapezoidal beam between 50% decrement levels. The exposure times at the angles shown are proportional to the length of the radial lines.

shown in Figure 7(a) and 7(b). Figure 7(a) shows the pattern that results when a beam with a trapezoidal profile makes one complete transit during one rotation of the gantry. In this example, the gantry was at 0° when the leading edge of the beam encountered the plane of interest and just reached 360° as the trailing edge arrived at the plane. The distribution is clearly weighted towards the 180° position. Figure 7(b) shows the weighting pattern when one complete gantry rotation coincides with the passage of the same beam from one 50% level to the next, i.e., the passage of a "geometric" beam as conventionally defined.

3.4. Combining Axial Exposure-Time Profiles

The ETP shown in Figure 2 may be considered as representing one step or "control vector" in the execution of a more complex treatment plan. A number of ETPs may be combined as shown in Figure 8 to show the effect of a number of vectors over the length of the tumor. The $\Delta ET(V_1)$ represents exposure time at various positions along the tumor axis for the first control vector (V_1). At any point at position z along the axis, the aggregate exposure time ET is found by summing the separate components

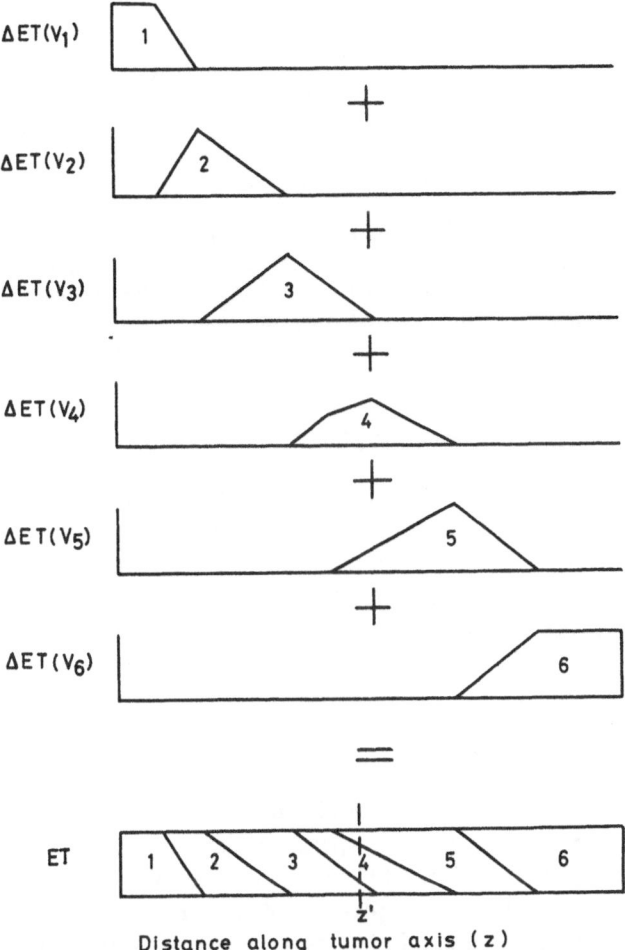

Figure 8. An axial exposure-time profile produced by a sequence of six "ideal" beams moving along a phantom. The numbers refer to the appropriate set of machine control data (control vectors) used by the treatment machine at various positions along the tumor axis.

$$ET_Z = \sum_1^N \Delta ET(V_N, Z)$$

It will be seen in Figure 8 that points in the plane z' receive dose contributions from vectors 3, 4, and 5 in the proportions indicated. The lines separating components from the different vectors represent the movement of the beam's collimator edges along the tumor with respect to time. Contributions from the various vectors may be changed while maintaining the same overall ETP as shown in Figure 9.

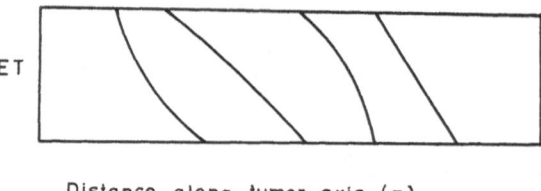

Distance along tumor axis (z)

Figure 9. Diagram illustrating that the contributions from various control vectors may be adjusted to produce different exposure weightings along the tumor axis while maintaining a combined profile as in Figure 8. The leading and trailing edges of the collimator blades are made to move at varying rates to contol the dose weighting from each vector.

3.5. Combining Tracks Using Exposure-Time Profiles

Figure 10 shows how the ETPs of three different tracks directed into the patient from different angles may be combined to produce a uniform exposure time over the length of the tumor. The time or exposure weighting may be nonuniform along each of the three contributing tracks if this is required to protect sensitive organs, such as the kidneys.

3.6. Combining Axial Exposure-Time Profiles and Transverse-Plane Exposure-Time Profiles

Axial and transverse-plane ETPs may be combined to represent the effect of a set of complex machine operations involving axial couch travel and gantry movement. The axial ETP, radial exposure weighting, and the profile of the beam in the transverse plane at Z' (Figure 8) may be combined to produce a transverse plane ETP. Consider a circular target volume 6.0 cm in diameter being irradiated by a 6.0-cm-wide ideal beam by a tracking technique. The section at Z' passes through the source rotation plane as the source arcs around the patient. As Figure 8 shows, the plane receives dose contributions from the treatment beam as the unit performs according to the machine control vectors V_3, V_4, and V_5 (each vector representing a distinct phase in executing the treatment as specified in the patient's treatment file).

Suppose that the total exposure time is 2.00 min with vectors, 3, 4, and 5 contributing 0.267 min, 1.200 min, and 0.533 min, respectively.

$$ET_{Z'} = \Delta ET(V_3, Z') + \Delta ET(V_4, Z') + \Delta ET(V_5, Z')$$

$$2.00 = (0.267 + 1.200 + 0.533)\,\text{min}$$

Suppose that at the start of V_3, the leading edge of the beam reaches plane $Z - Z'$ as the gantry reaches $0°$ while moving clockwise at $120°\,\text{min}^{-1}$. After

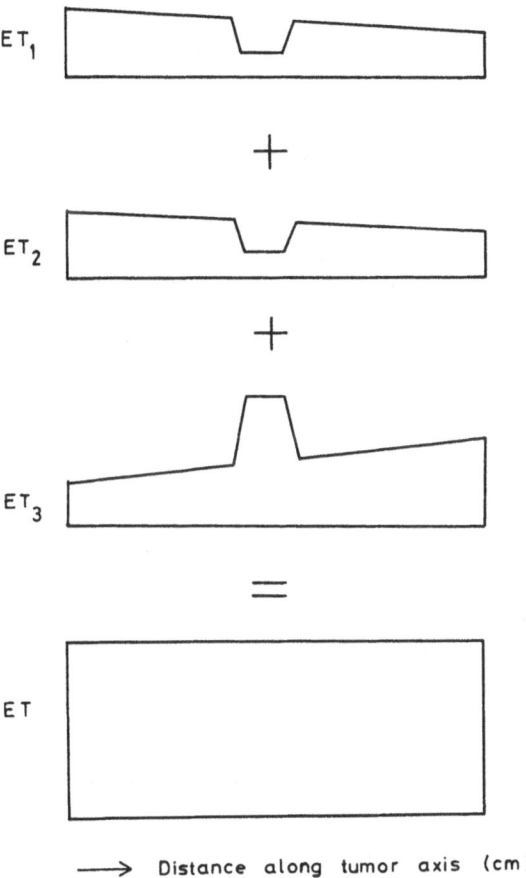

Figure 10. Combining exposure-time profiles for a number of tracks to control the dose contributions for different angles of beam entry along the tumor axis.

0.267 min, the program moves on to V_4, which specifies a gantry of speed 180° $\min^{-1}$. After an additional 1.200′ min, the program moves on to V_5, which requires the gantry to move at 240° $\min^{-1}$.

It is seen that V_3 applies for 32°, V_4 applies for a further 216° until the gantry is at 248°, and V_5 applies for an additional 96° until the gantry is at 360° + 16°. The times taken for the gantry to pass through 20° intervals (stations) are easily calculated and represented as a radial time-weighting diagram with the length of the lines proportional to the exposure time as indicated in Figures 6 and 7. Such a diagram clearly takes no account of the beam width, i.e., the beam dimension in the transverse plane. The transverse-plane ETP may be calculated and used to represent both the gantry speed and the effect of beam width in much the same way that axial ETPs represent

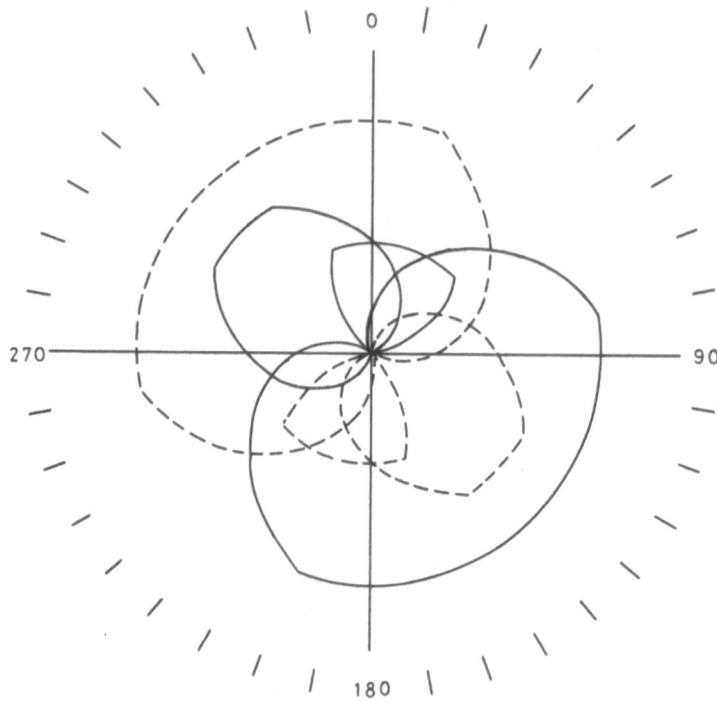

Figure 11. Radial exposure-time profiles for the transverse plane at z' irradiated by control vectors 3, 4, and 5 as shown in Figure 8. The figure shows the three profiles separately (solid lines) and also the exit-side contributions (broken lines) around a circular locus of radius 3.9 cm and concentric with a 6.0-cm diam. target volume that is being irradiated by a 6.0-cm wide beam.

the couch speed and beam length in the axial direction. Figure 11 shows the ETP profile for each of the vectors 3, 4, and 5 around the circumference of a circle of 3.9-cm radius and concentric with the 6.0-cm diameter target volume mentioned previously, the beam width being 6.0 cm and thus subtending an angle of 100° at the centre of rotation. Figure 12 shows the cumulative effect of the three vectors. These representations of the effect of the treatment beam in a single transverse plane may also be displayed in Cartesian form.

     These descriptions of the operations of the treatment unit are obviously very simplified, and allowances would have to be made for the effect of the beam profile in both the axial and transverse directions for the more complex geometry if the ETP were required around a noncircular locus or if the treatment involved vertical and lateral couch movements and if cross contributions are to be taken into account. For optimization purposes, a simplified beam model will suffice, but for a final accurate three-dimensional dose calculation, more sophisticated models must be used.

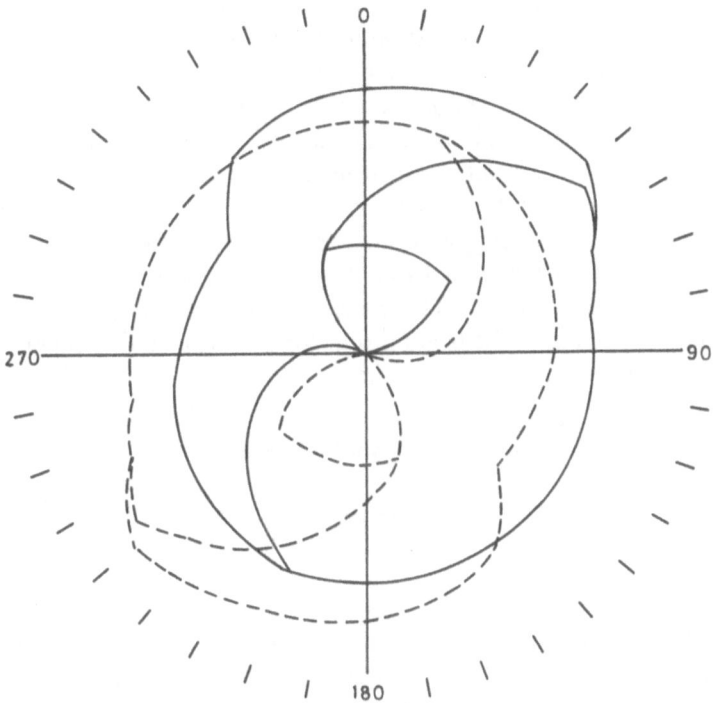

Figure 12. Summated radial exposure-time profiles for the transverse plane at z' that is irradiated under the same conditions as in Figure 11.

4. CONTROLLING RADIOTHERAPY DOSE DISTRIBUTIONS IN THREE DIMENSIONS USING A COMPUTER-CONTROLLED TRACKING UNIT

4.1. Slice-by-Slice or Field-by-Field Treatment and Planning

In order to control the dose distribution in three dimensions, both the shape of the high-dose zone and the dose distribution within it must be controlled. If the dose distribution is correct for each body cross section, it will be correct for the whole volume. Similarly, if the shape of each radiation field and the dose distribution across the field at the tumor depth are correct for each gantry angle when a cross-fire technique is used, the distribution will then be correct for each body cross section provided that the beam energy is high enough. These two approaches lead to the development of a number of treatment methods that differ as much in the approach to treatment planning as in their execution. Present-day treatment-planning systems are quite effective at calculating dose distributions in the transverse plane with corrections made for cross contributions from beams having their central axes displaced from the calculation plane in the superior–inferior axis. Provided that the beam model represents the dose distribution with a sufficient accuracy over the length of the tumour, three-dimensional dose distributions can be computed, albeit inconveniently and slowly at present. For treatments in which the volume is built up slice by slice with the couch remaining stationary during exposure, the exposure timer or transit dose-meter performs its conventional role. For treatments in which the machine isocenter is displaced relative to the patient during exposure, it is necessary to integrate the dose under the beam profile for each calculation point, and the timer performs a trivial role.

4.2. Controlling the Shape of the High-Dose Volume

4.2.1. CONTROLLING THE SHAPE OF THE HIGH-DOSE VOLUME IN THE TRANSVERSE PLANE

Many textbooks deal with single-plane treatment planning for one target volume of simple shape in the transverse plane. The basic policy is to decide on the use of a cross-fire technique with fixed fields or to use an arcing or rotation technique. It is often claimed that arcing treatments lead to high integral doses, and this may well be the case using conventional manually controlled equipment. In such treatments, the beam size may simply be set at the minimum required to cover the widest part of the tumor, and the target volume is thus enclosed in a high-dose volume conforming to a simple solid of revolution around a horizontal axis. When computer-controlled machines are used, both the collimator angle and field size can be adjusted continuously to give the best match between the projected target outline and the beam for all gantry angles. When automatic machines are used with suitably designed treatment plans, there will be little difference between the irradiation high-dose volumes and integrated doses whether stationary or moving-beam treatments are employed. A major planning consideration is the disposition of the unavoidable dose delivered to the healthy tissue outside the target volume. Arcing and rotation techniques spread the unwanted dose uniformly around the target volume, and some therapists prefer such treatments, especially for the pelvic regions because they avoid "pinch spots" at beam intersections. If, however, it is required to reduce the dose preferentially to sensitive structures outside the target volume, then fixed-field techniques may be preferred. Figure 13 shows the geometrical fit for some simple-target volumes irradiated using a parallel pair, a three-field technique, a four-field technique, and a five-field technique. The lack of control over the irradiated volume with a parallel pair is clear. The three-field plan provides a good fit

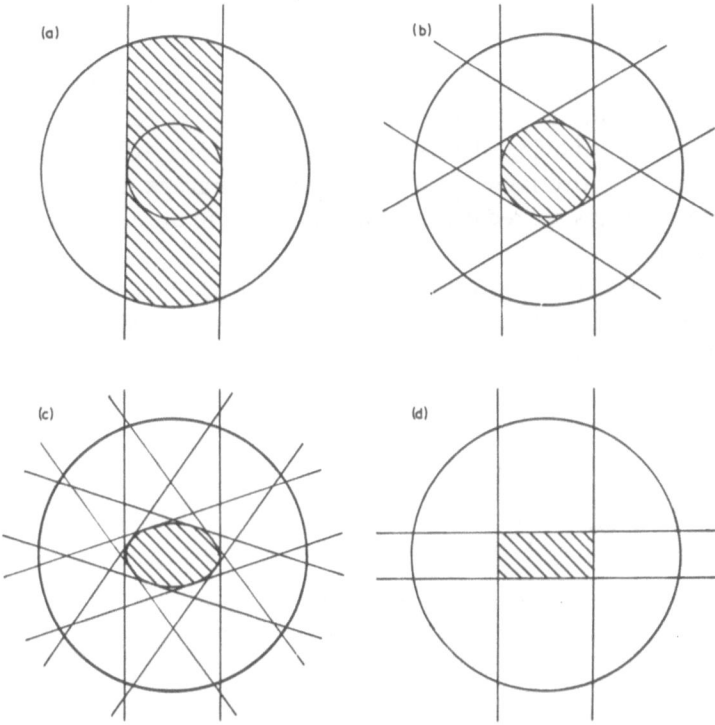

Figure 13. Diagram illustrating the possible closeness of the geometrical fit between simple target volumes and the high-dose volume in a single plane using various beam confirmations. (a) A circular-section tumor treated with a parallel opposed pair of beams. (b) A circular-section tumor treated with three beams with 120° angles between them. (c) An elliptical-section tumor treated with five beams with 72° angles between them. (d) A rectangular-section tumor treated with a four-field "box".

for a circular target volume, and a five-field plan provides a good fit for the eliptical-section target. A four-field "brick" provides a perfect fit for the rectangular-section target volume. Much more complicated target volumes (including two targets in the same cross section) can be irradiated if complex treatment plans are used. The synchronous shielding techniques described in the first section of this chapter indicate the possibilities. With computer-controlled machines, treatment plans must be devised to use the existing collimated beam. Chin *et al.*[17] show some interesting and complex distributions for the irradiation of the pelvic and thoracic regions. Case studies in the next section show the irradiation of pelvic nodes using a track with multiple overlaid arcs on two isocenter loci. A preliminary study (Section 7.4) at the RFH shows how crescent-shaped dose distributions may be produced in the transverse plane using the MS90 tracking unit.

4.2.2. SHAPING A COMPOSITE FIELD USING A TRACKING UNIT

Figure 14(a), 14(b), and 14(c) illustrates how a composite field is built up by simply abutting a number of small beams, overlapping a number of small beams, or moving the beam along the target volume during the exposure. When overlapping or continuously tracked beams are used, corrective measures are required to avoid a fall-off in dose at the ends of the target volume.

In principle, beams may be abutted or overlapped manually, but this is not very practical because of the time it takes and the risk of errors in setting collimator sizes and positioning the beams. This is, however, a simple and

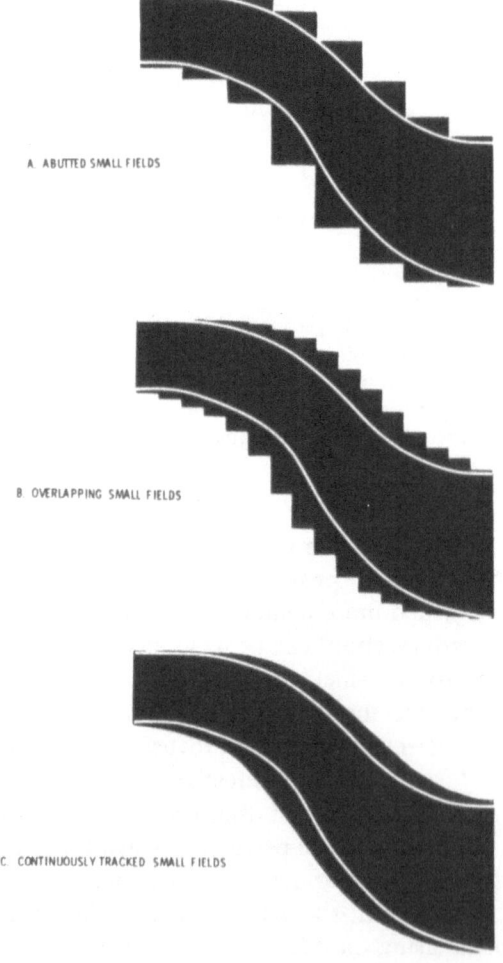

A. ABUTTED SMALL FIELDS

B. OVERLAPPING SMALL FIELDS

C. CONTINUOUSLY TRACKED SMALL FIELDS

Figure 14. Shaping an irradiated field using a computer-controlled tracking machine. (a) Using abutted small beams. (b) Using overlapping small beams. (c) Using continuously tracked small beams.

practical procedure using computer-controlled machines, although using simple abutted fields is not recommended for use with linacs because of the sharp beam edge. It may be noted, however, that penumbra may be artificially produced by opening or closing the collimator blades slightly under computer control—a process that may be termed dynamic penumbra generation (DPG).

It will be appreciated that the target outline indicated in Figure 14 may be considered as a projection of the target volume or simply as the shape of the field required in a particular plane in the tumor. The rectangular areas may thus represent a series of high-dose slices or a number of beam cross sections.

4.3. Controlling the Dose Distribution along the Tumor Axis

The dose delivered to a point in a patient or a phantom will depend on the time the point remains in the beam and on the dose-rate profile of the beam.

Considering the integrated dose profile at a depth d in a phantom that is being irradiated by an ideal beam of constant cross section, traveling at a constant speed τ shows that there is a region of uniform dose over the central region and that the dose falls off linearly over a distance equal to the beam length (X) at each end of the track as shown in Figure 2. The dose delivered to a point P in the phantom is found by integrating the dose rate with respect to time.

$$D_P = \int_{t_1}^{t_2} R_P \, dt \tag{1}$$

where D_P is the total dose delivered to the point P, R_P is the dose rate at P at time t, and t_1 and t_2 are the times corresponding to the beginning and end of the exposure. The dose rate at P is given by

$$R_p = R_{air} T(d) \rho(x) \tag{2}$$

where R_{air} is the in-air dose rate, $T(d)$ is the tissue–air ratio at depth d, and $\rho(X)$ is the off-center ratio at distance x from the central axis at the same distance from the source as P.[18] If P is being translated through the beam at constant speed τ,

$$D_P = R_{air} T(d) \frac{1}{\tau} \int_{x_2}^{x_1} \rho(x) \, dx \tag{3}$$

where x_1 and x_2 are the displacements of P from the central ray at times t_1 and t_2. This may be expressed as

$$D_p = R_{air} T(d) X_E / \tau \tag{4}$$

where X_E is the effective length of an ideal beam as discussed in Section 3.1 of this chapter.

As stated, a first approximation of X_E may be taken as equal to the 50% length at the depth of point P. In an actual treatment, it will be necessary to integrate the area under a fairly complex dose-rate profile.[19, 20]

Since $D_p = R_{air}T(d)X_E/\tau$, the dose level is controlled by adjusting X_E/τ. Increasing X_E or decreasing τ will increase the dose delivered; decreasing X_E and increasing τ will decrease the dose delivered. The ratio can be adjusted as the beam travels along the tumor axis to compensate for variations in tumor depth and the presence of inhomogeneities, such as bone or lung. Allowances must be made for variation of the in-air dose rate (R_{air}) and the tissue-air ratio [$T(d)$] with beam size.

Since the speed of couch travel (τ) controls overall treatment time, it is desirable for this speed to be as high as possible. This, however, would mean increasing X_E if the dose were to be kept constant. If the tumor is of uniform width, there is no reason why longer beams should not be used. The procedure for selecting various combinations of beam lengths and tracking speeds is illustrated in the section on axial dose profiles and the examples in Section 7.

If the tumor has an irregular outline, other considerations apply, and it may be necessary to strike a compromise between the closeness of fit to be achieved between the target cross section and the irradiated field, and the total exposure time.

4.4. The End-of-Track Technique

Returning to the dose profile in Figure 2, it is clear that a method is needed to produce a sharper fall-off in dose at the ends of the treatment volume. The easiest way to achieve this for a simple linear track is to use lead shielding blocks at either end of the treatment volume.

If an ideal beam of constant width and length exhibiting negligible beam divergence travels at constant speed over the surface of the phantom, Figure 15(a) represents the situation where the beam starts over one lead block and then travels along the phantom until it finally comes to rest on a lead block at the other end of the treatment volume. The exposed portion of the beam is shown as a solid block and a number of features are apparent. The leading edge of the beam advances at speed τ, and the trailing edge (defined by the lead block) will remain stationary until the leading edge has moved a distance X and thus exposed the whole beam. The center of the exposed part of the beam thus travels at a speed $\tau/2$ until the beam is fully exposed. Once the beam is clear of the block, it moves forward at speed τ until the leading edge of the beam reaches the distal shielding block. The pattern of exposure that occurred at the beginning of the track is now repeated in reverse. Figure 15(b)

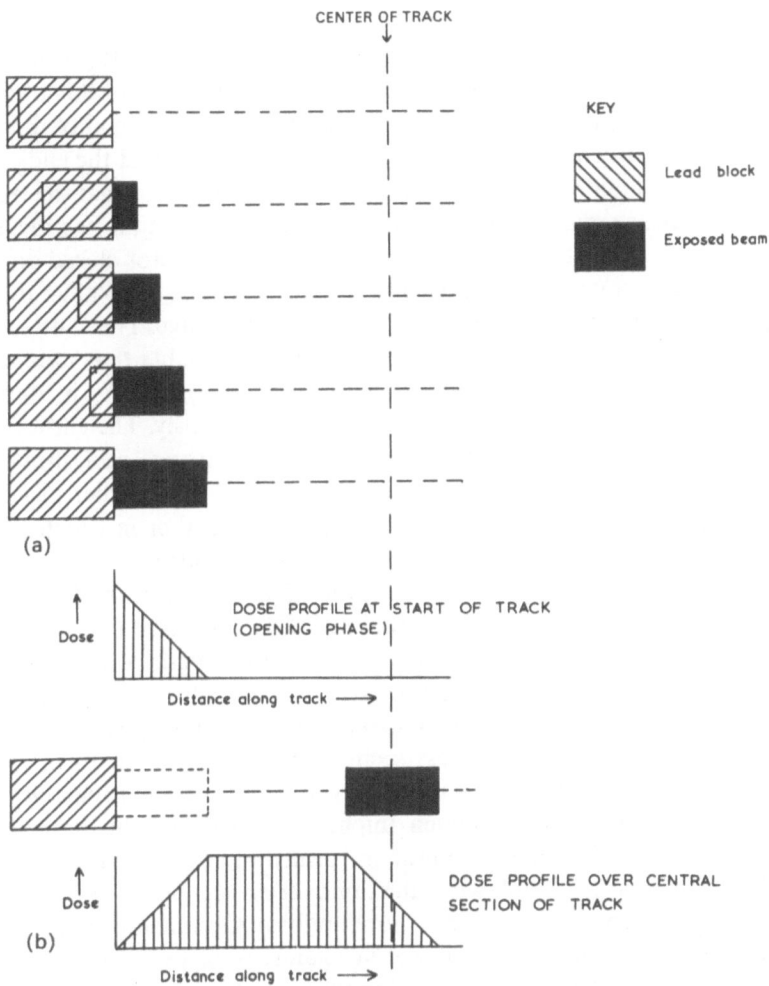

Figure 15. Diagrams showing the pattern of beam exposure using lead shielding blocks or a tracking technique using collimators capable of giving zero beam length. (a) Pattern of beam exposure as the beam moves off a lead shielding block and travels along the tumor axis at constant speed (upper half of figure). (b) Exposure profile resulting from the track in (a) (lower half of figure).

shows dose profiles at the ends of the track and over the central portion. These profiles may be combined as shown in Figure 8. It is clear that these distributions may be simulated using a collimator that can produce beam lengths from 0 to X and a treatment table that can be positioned accurately with respect to the beam central axis. This requires using a computer-controlled machine if the technique is to be practical for patient treatment.

It is, in practice, impossible to produce a zero-length beam because of penumbra and mechanical design limitations. Fortunately, the desired uniform dose profile can still be produced as shown in Figure 16(a) and 16(b); this method is in routine at the RFH. The essential features of this track are that the minimum beam length is used at the beginning and end of the track and also over the central portion. To obtain a sharp cut-off at the ends of the volume, the collimator is opened at twice the couch speed, so that the trailing edge of the beam remains stationary. When the beam length has increased from X_{min} to $2.X_{min}$, the shutter is closed, and the collimator closed down to X_{min} again. The couch is then advanced by a distance $X_{min}/2$ to align the new trailing edge with the position of the original leading edge. The shutter is re-opened and the couch advanced at a speed that is double that used in the opening phase. The speed τ is calculated using Eq. (4), and X_E is found by integrating under the beam profile, as indicated previously. The end-of-track technique using a shutter closure with a couch advance and shutter re-opening was devised by J. A. Brace, and it supersedes an earlier procedure based on the same principle and due to the present author in which a rapid couch advance was used with the shutter remaining open.

For tumors with convex ends, using the standard end-of-track technique can lead to excessive areas being irradiated if X_{min} is large (say, 4 cm). An alternative approach would involve extending the length of the track and using the minimum available beam length (which might be less than 1.0 cm on a linac) throughout the track and so obtain the closest possible fit of the composite field to the target cross section. In addition to delivering some unwanted dose beyond the ends of the target volume, this approach leads to protracted treatment times, which might inconvenience the patient and increase leakage radiation contributions. Ideally, the beam length X_E should be adjusted so that it is small when the tumor cross section is varying rapidly and lengthened when the outline is more uniform. Optimizing combinations of X_E and τ to balance the degree of fit against total treatment time is an important consideration when planning these treatments.

5. A NOTE ON TREATMENT-PLANNING STRATEGY

5.1. Thin-Slice and Thick-Slice Planning

There are two basic approaches that may be adopted in three-dimensional treatment planning. In the thin-slice approach, the initial beam configurations and dose weightings are selected on the assumption that the treatment unit has unlimited resolution in the axial direction. It is assumed that plans may be devised independently for a series of transverse body sections and the dose distribution for each plane results from only the plan

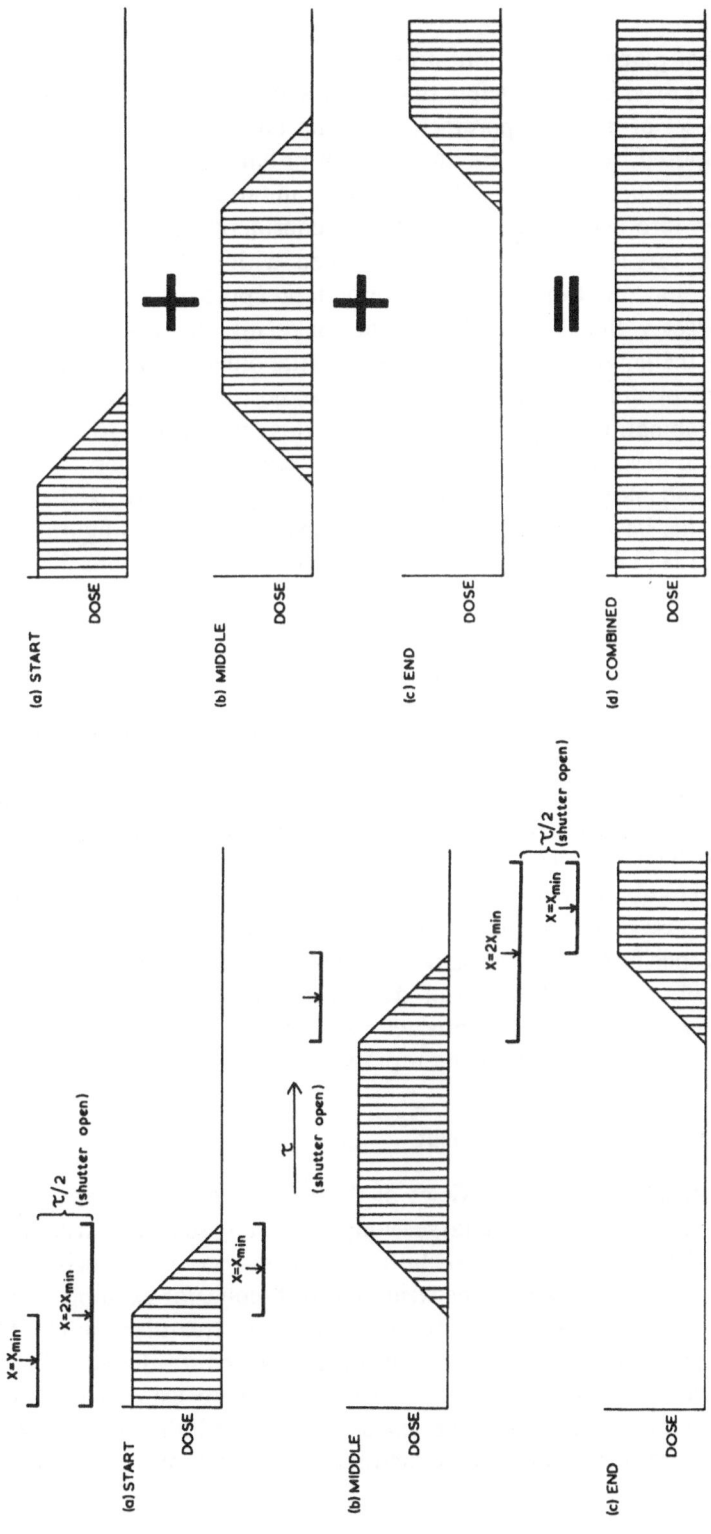

Figure 16. (a) Diagram showing the pattern of collimator opening and tracking speed to produce a uniform dose distribution along the tumor axis when the minimum attainable beam size is greater than zero. For this "high conformity" track, the minimum attainable beam length X_{min} is used over the whole track except for the end of the track correction phase. Dose profiles corresponding to the collimator opening phase and over the central section are shown separately and combined. (b) Combining the axial-dose profiles for the beginning, middle, and end phases of a track to obtain a uniform dose distribution along the tumor axis.

devised for that plane. The optimization process may therefore be directed toward finding machine parameters that provide the best approximation of this idealized set of machine parameters, taking into account the limits in resolution imposed by using a treatment beam of finite length.

In the thick-slice approach, the practical limitations of the treatment unit are taken into account at the outset by considering the whole of the section of the target volume that is contained within the length of the treatment beam at one time. This more pragmatic approach requires using composite target volumes that result from coalescing all those transverse planes within the length of the beam. The essential difference between thin- and thick-slice planning is that in the first case, no assumptions are made about the combinations of beam lengths and dose weightings to be used along the tumor axis, whereas in the second case, certain parameters, such as beam lengths and effectively beam widths and isocenter locus, are fixed at the outset. Once this basic strategy has been decided, the beam configurations can be selected field by field or slice by slice as discussed in Sections 5.2 and 5.3.

5.2. Slice-by-Slice Treatment Planning

Each thin slice or composite thick slice can be entered separately into a treatment-planning system and a suitable plan devised in the conventional manner. These plans are then combined in the treatment-planning system to produce a three-dimensional dose distribution that can be achieved by the treatment unit. The first step in this process is to balance the dose weighting from each plan to obtain a uniform dose profile along the tumor axis but with the dominant dose distribution for each calculation plane coming from the plan designed specifically for that plane. It will be apparent that although it is advisable to obtain a uniform axial dose distribution, this may not be sufficient to ensure that doses are satisfactory at points off the axis, especially if the tumor cross section is changing rapidly along the length of the tumor.

5.3. Field-by-Field Treatment Planning

Using a series of transverse views of the patient obtained from a CT scanner, the clinician outlines the target volume and any sensitive structures and specifies the required dose distribution in these regions. A decision is usually made at this stage about the treatment technique, for example, for a spinal tumor a tracked wedged pair or a three-field treatment might be selected as outlined in Section 6.2. If the tumor is to be irradiated using a number of fields (each field being a composite built up using a tracking procedure), the beam entry angles are chosen, and projections of the target volume as seen from the source are constructed. For each of these fields or tracks, the beam length—tracking-speed pattern is chosen as described in Section 4.4. Typically, a standard pattern will be used. This basic tracking pattern is then modified using the tracking speed with the aid of exposure time and ADP to give the required dose weighting along the tumor axis. The beam widths can be taken directly from the projected target profile with corrections made to allow for the influence of other fields in the manner described by Perry et al.[6] For standard techniques, these corrections can easily be determined by studying transverse-plane treatment plans on a conventional planning system.

It is therefore necessary to observe dose distributions over the whole target volume, and they will have to be computed or measured. An alternative procedure would be observing dose distributions as profiles along a number of strategically chosen loci running in the superior–inferior axes as indicated previously in Figure 5.

It will be apparent from the sections that follow that the methods currently used at the RFH are based mainly or entirely on the empirical thick-slice approach except where radial time or exposure patterns are used for testing arc tracks (Section 7.1). It is anticipated, however, that the thin-slice method may provide the basis of a more logical approach to treatment planning.

6. BASIC TREATMENT METHODS

6.1. Machine Operational Modes

Conventional treatment units may be operated in two basic modes; viz., with the beam stationary during exposure or with the beam moving during exposure. Computer-controlled tracking units may be operated in four basic modes, viz.,

1. Mode A: moving couch–moving gantry.
2. Mode B: moving couch–stationary gantry.
3. Mode C: stationary couch–moving gantry.
4. Mode D: stationary couch–stationary gantry.

In modes A and B, the couch moves relative to the machine isocenter so that the isocenter follows either an axial locus or a locus in a transverse plane in the patient. Some basic treatment methods using these modes are outlined in the next section.

For treatments using modes A and B, it is not yet practical to use commercial treatment-planning systems (TPSs) to calculate dose distributions. For modes C and D, conventional TPSs can calculate dose distributions, although it can take hours of planning time because of the large number of beams and calculation points. These problems should soon be overcome by more powerful planning systems with facilities for moving automatically from beam to beam and point to point.

6.2. Some Basic Treatment Techniques

6.2.1. TYPE A TRACKING TREATMENTS—THE ARC TRACK

For type A tracking treatments, an arcing technique is combined with a moving-table technique. This is a development of the simple arcing procedure carried out on the old Marconi unit,[22] where it was used mainly to achieve an adequate depth dose. This procedure has been retained and improved on the new cobalt unit by the closeness of fit that can be achieved between high-dose volume and target volume with an absence of high-dose regions outside the target volume. The improved patient tolerance to arcing treatments when large volumes are irradiated is an important reason for its use despite difficulties in treatment planning and machine control that are encountered in the technique. Although modern control systems are able to execute arc tracks successfully, improved treatment-planning procedures are needed to optimize the plan efficiently.

The patient is passed through the source rotation plane while the source is made to arc around the patient's long (superior–inferior) axis. Throughout the exposure, the patient is continuously repositioned vertically and laterally so that the iso center follows (tracks) along a predetermined locus inside the patient. The shape, size, and orientation of the high-dose zone in each section is determined by a combination of beam size, arcing angle, and arcing speed. The position of the high-dose zone within the body outline is adjusted by the couch's vertical and lateral displacements. These displacements are specified relative to the intersection of a vertical and a horizontal reference plane, with the line of intersection coincident with the gantry rotation axis. The dose delivered to a point passing through the beam depends on how long it remains in the beam, so it is controlled by a combination of tracking speed and beam size in the direction of longitudinal movement. Figure 17 shows how various parameters are used to control dose distribution in the patient.

6.2.2. TYPE B TRACKING TREATMENTS—LINEAR TRACKS

In type B tracking treatments, a number of fields are used in a cross-fire technique. Each field is a composite built up by passing the patient through the source rotation plane with the source stationary. The beam size is adjusted automatically according to the tumor projection for the body section currently in the beam. This process is repeated for as many beams as required. A single field may suffice if the spine is to be treated. For each field,

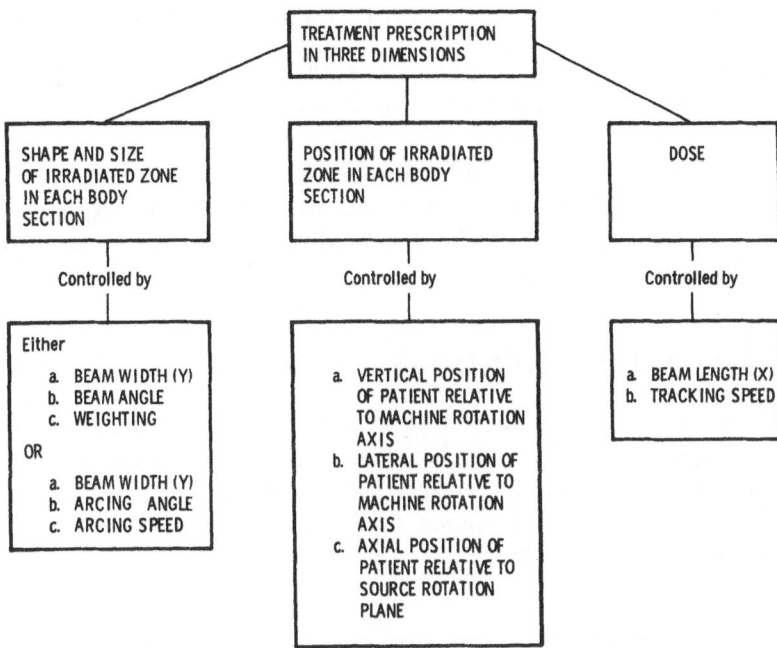

Figure 17. Machine parameters used to control the dose distribution in three dimensions in the tracking technique.

the position of the machine isocenter is again controlled by adjusting the couch vertical and lateral positions. As in type A treatments, the dose delivered is controlled by a combination of couch longitudinal speed and beam length. This and the previous technique differ in that the dose weighting at each gantry angle is controlled by the arcing speed in type A treatments but in type B all weighting is done by the longitudinal speed and beam lengths, and this can be different for each field.

6.2.3. TYPE C TREATMENTS—ABUTTED SECTIONS

Type C treatments are a discontinuous form of tracking treatment in which the automatic positioning facility is used to treat tumors section by section. Each section is abutted accurately to its neighbour to avoid overdosing or underdosing at the interface. In principle, this technique can be carried out manually. However, in practice it would be impossibly time consuming, and there would be a significant risk of errors arising from failure to correctly reset machine parameters. Type C treatments in this simple form are not recommended for use with linacs because of the sharp beam edge, as discussed previously.

6.2.4. TYPE D TREATMENTS—OVERLAPPING SECTIONS

In type D treatments an attempt is made to achieve a closer fit to the tumor volume while reducing the risk of dose inhomogeneities at section interfaces by overlapping consecutive treatment sections. Again, this can be done safely and efficiently only under computer control. This is a very promising technique that can be used on cobalt units and linacs, and it has the useful advantage that standard treatment-planning systems can calculate the resultant dose distributions.

6.2.5. TYPE E TREATMENTS—TRANSVERSE-PLANE TRACKS

It is sometimes desirable to treat a curved surface by using a number of small tangential beams. The couch can be made to follow a curved locus in the source rotation plane to produce such an effect. This method can be combined with the abutted section or overlapping section method to fit surfaces curved in two planes.

6.2.6. TYPE F TREATMENTS—BIFURCATED TRACKS

For some tumors, such as those affecting the pelvic and para-aortic lymph nodes, the tumor divides from one volume into two branches. Such tumors are usually treated on conventional machines using a mantle technique where two parallel opposed beams are shaped using specially fabricated lead blocks. A tracking technique using two loci for the isocenter achieves an effective treatment that reduces the high-dose volume considerably. Typically, a bifurcated track would be used in combination with the type D treatment method as illustrated in Section 7.3.

7. SOME EXAMPLES OF PHYSICS PLANNING PROCEDURES

7.1. Planning an Arc Track

An arc track is a hybrid treatment in which couch translation is combined with gantry rotation. The usual procedure for planning such treatments is to find a suitable single-plane plan for each section, using a conventional planning system on the assumption that an integral number of arcs occur as the plane of interest passes through the source rotation plane; i.e., it is assumed that it is treated with one complete arc per track. This single-plane plan provides the mean TAR or TMR with appropriate weighting for arcing speeds. The tracking speed is then calculated from

$$D_p = \text{Dose rate} \times \text{time}$$

$$= D_{\text{air}} T(d) \frac{X_E}{\tau}$$

Exposure-time profiles are then constructed over the length of the tumor as shown in Figure 18. The variation of X_E along the track is usually in accordance with some standard pattern, such as the high-resolution track described in Section 4.4. From the ETP, the ADP is constructed by multiplying various components of the time profile by appropriate dose rates, suitably corrected for TAR with allowance for field size

$$\Delta D_p = \Delta ET[R_{\text{air}}]T(d)$$

It will be noted that the effects of penumbra and scatter are taken into account at this stage by the choice of X_E. In practice, it is not easy to set the exact equivalent of one arc per traverse of the useful beam; fortunately, arcing treatments are fairly tolerant of departures from exact dose weighting. Nonetheless, it is necessary to ensure that the actual weighting pattern does not depart too far from that intended. Two precautions are therefore taken. The treatment plan is divided in half, with each part intended to deliver an identical dose to an identical target volume. However, each track is a mirror

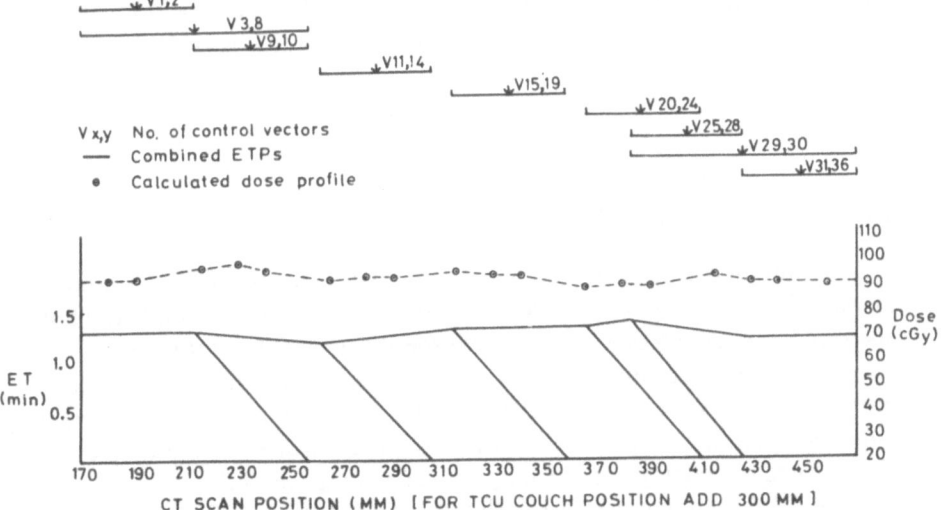

Figure 18. Diagram showing the construction of the exposure-time profile and the calculated dose profile along the isocenter locus of an "arc track".

image of the other as far as the arcing angles are concerned. For example, the gantry may be positioned at 240° at the beginning of the first half of the treatment and caused to move in a clockwise direction. For the second half, the gantry will start at 120° and move in a counterclockwise direction. While these measures ensure symmetry, they do not ensure the equivalence of an integral number of arcs. A radial dose-weighting pattern is therefore calculated at a number of cross sections and compared with that used for planning purposes. If the patterns look significantly different, the actual weights can be entered into the planning computer to check the distribution and the effect on the axial dose rate. If necessary, the machine arcing speeds can be adjusted over the length of the track to achieve a better agreement between the intended and the actual radial dose-weighting pattern. The overall dose distribution may then be confirmed by calculation or measurement. It will be noted that unless a sophisticated calculation of X_E is performed over the length of the track, or, more precisely, unless the dose-rate–time integral is properly evaluated, the final dose will not be accurately known. It will be appreciated that balancing the treatment is based on machine operations rather than dose calculations, in the expectation that the required machine operation is known for each cross section. The final dose distribution may be displayed as a set of transverse-plane isodoses and an axial dose-distribution profile. Doses may be expressed in cGy or as percentages of some reference point, typically at the isocenter on a plane at or near the center of the tumor. Preparing a treatment file for the MS90 unit is done with the aid of the editor program. Couch and gantry speeds are entered into the treatment file, which then calculates the gantry angle reached at different positions along the tumor axis. Beam lengths and widths appropriate to these positions are then entered through the computer keyboard.

7.2. Planning a Multitrack Treatment

A multi field track is treated using a number of composite fields built up as described in Section 4.2.2. Again, standard basic beam-length tracking speed patterns are used, with the actual speed calculated as for the arc track.

The required beam sizes and dose weightings are again found for a number of transverse sections using a conventional planning computer. In multitrack treatments, all the dose weighting is controlled by the axial couch speed, since the gantry is stationary for each composite field.

An ETP is constructed for each track, and they are converted to ADP that may then be combined. Again, the key to the dosimetry is integration under the dose-rate profile along the track with respect to time. Figures 19(a)

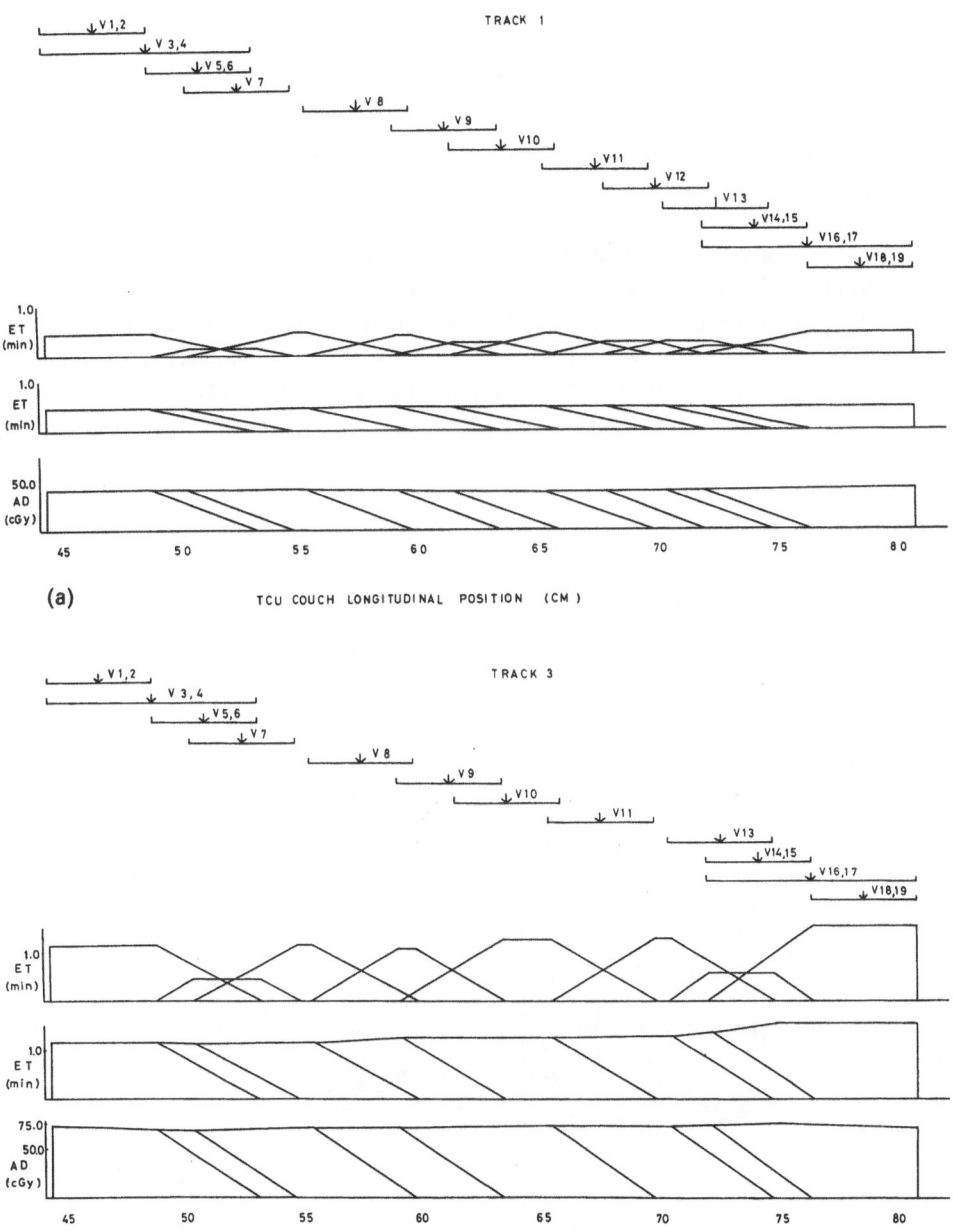

Figure 19. Diagrams showing the method of constructing exposure-time profiles for a three-track treatment. Three linear tracks were used to irradiate a spinal tumor; track 1 was a direct unwedged posterior field, while tracks 2 and 3 were oblique wedged fields, with the major dose contribution delivered by the oblique fields. (Track 2 is similar to track 3.) (a) Exposure time and dose profiles for track 1. (b) Exposure time and dose profiles for track 3.

and 19(b) show the ETPs for the direct posterior field and one of the two posterior oblique fields.

7.3. Planning a Bifurcated Track

The target volume required to enclose the pelvic and para-aortic nodes takes the form of an inverted Y. Such volumes are conventionally enclosed in simple brick-shaped volumes or in Y-shaped volumes produced by parallel opposed fields using the mantle technique. In either case, the high-dose volume will be much larger than the target volume. The high-dose volume may be considerably reduced by treating the tumor with a series of overlapping high-dose slices. Figure 20 shows a target volume reconstructed from 8-mm-thick CT scan slices taken at 16-mm intervals over the length of the tumor. Figure 21(a) and 21(b) show the AP and lateral projections of the target volume.

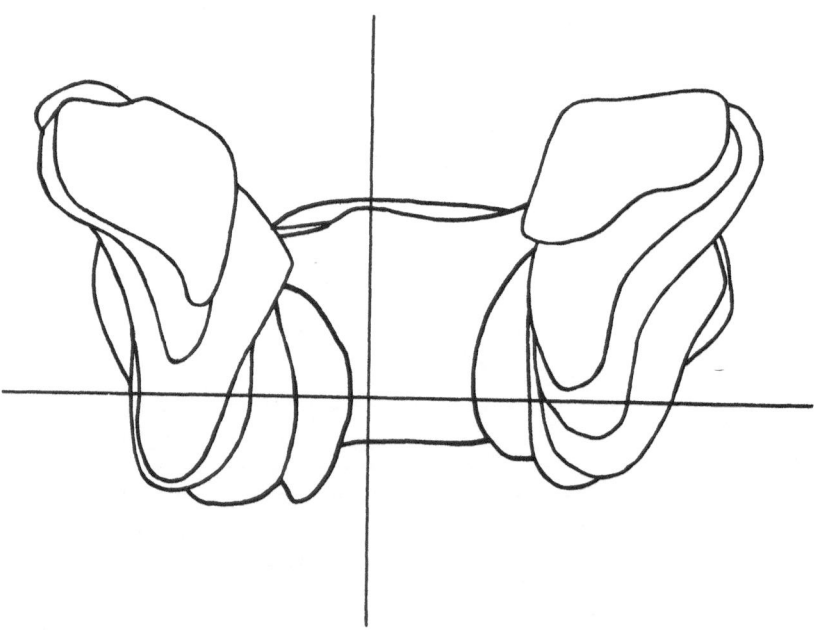

Figure 20. Pelvic-node target volume from a CT body scanner displayed at 16-mm intervals to aid treatment planning.

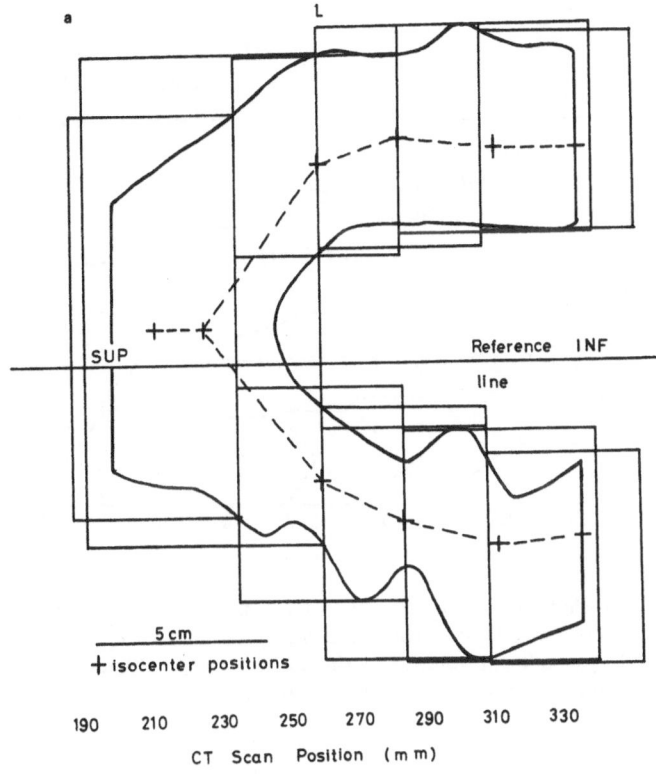

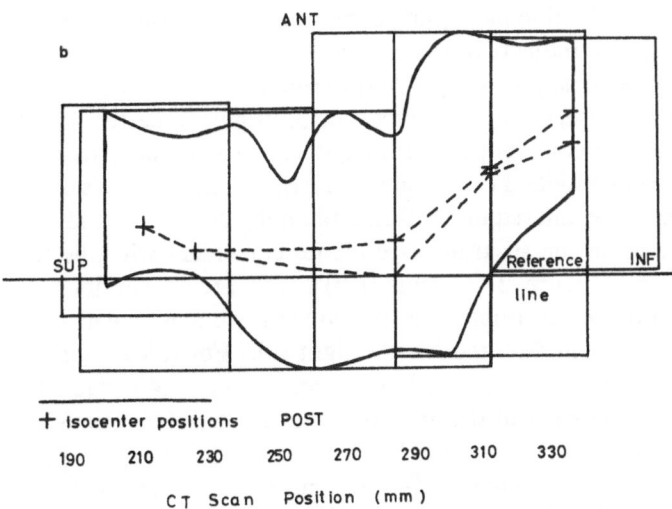

Figure 21. Projections of a pelvic-node target volume showing isocenter positions and a series of superimposed "thick-slice" target volumes. The reference lines are taken from CT scanner views. (a) AP projection. (b) Lateral projection.

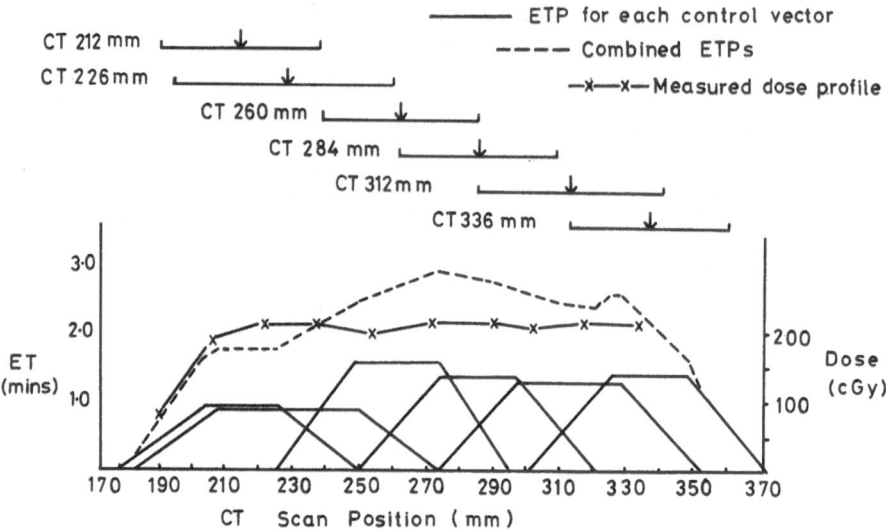

Figure 22. Diagram showing the overlay pattern for a series of arcing treatments used to irradiate a bifurcated pelvic-node target volume. In this representation, all the arcs in each plane are considered to be part of the treatment for a single target volume, although over the inferior part of the volume, there are two target volumes in each cross section.

A series of overlapping slices is superimposed onto the projected target volumes as shown; each slice is drawn wide enough to enclose all the tumor contained within the thickness (50% definition) of the slice. An axial overlay pattern is drawn to indicate the extent and magnitude of cross contributions along the tumor axis (Figure 22). A trapezoidal-beam profile (Figure 1) is used for this purpose, since, for overlaid fields, it gives a better impression of the effect of cross contributions from one slice to the next than does the simple ideal beam model. For the tumor illustrated, the patient's outlines and target volume were entered into a Rad-8 planning computer, and a plan was devised for each of the six transverse planes coincident with the six source rotation planes [Figure 21(a) and 21(b)]. The arcing configurations are shown in Figure 23. The tumor was treated with two files, one irradiating the left side of the patient and another the right side. For calculation purposes, the two superior arcs were each regarded as one 240° arc. The dose contributions from each of the arcs to points at the intersections of the six transverse planes and the isocenter loci were calculated using the products of the axial and transverse-plane off-center ratios. The dose weightings from each of the six (pairs of) arcing configurations were then empirically adjusted to produce an acceptable dose distribution along the isocenter locus. The aggregated dose distributions from all the arcs were then calculated for the

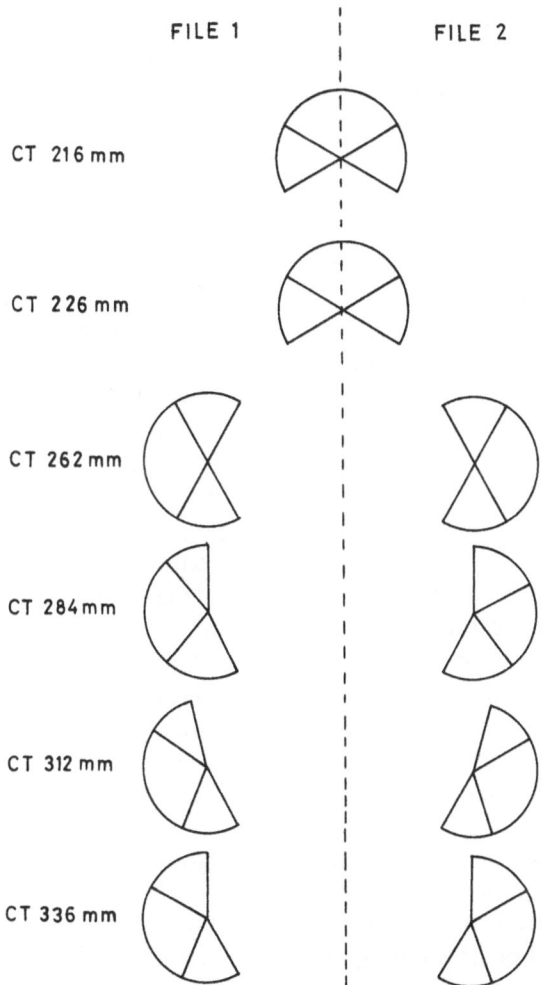

Figure 23. Arcing configurations used to irradiate a bifurcated pelvic-node tumor. For treatment purposes, the two most superior arcs were split into two symmetrical halves and two treatment files prepared, one for the right side of the patient and one for the left side. Both start from the same superior setting-up mark.

original transverse planes and for some intermediate planes using the Rad-8 system "summation of plans" program. Cross contributions from distant planes were added as uniform (small) doses over the whole plane, since the beam profile degenerates to a low-level (nearly) uniform dose due to leakage and scatter.

The calculation process is very tedious and slow using conventional planning systems. It is anticipated, however, that newer planning systems will process such plans automatically and quickly.

Typical distributions for three planes are shown in Figure 24(a), 24(b), and 24(c). It will be appreciated that each plane may be considered as being treated with every arc, and the only significant difference between them lies in the relative dose contributions as determined by the beam profiles in the axial direction.

7.4. A Note on the Transverse-Plane Track

Crescent-shaped high-dose cross sections may be useful for treating tumors around the chest wall. Techniques for producing such irradiated volumes have been described by Proimos et al.,[23] who used a synchronous shielding technique. Chung-Bin et al.[24] described a treatment for mesothelioma of the pleura using a computer-controlled Varian Clinac-4. This unit positions the machine isocenter in the body outline and adjusts the gantry angle in accordance with a stored program in much the same way as the MS90 CCTCU at the RFH. The isocenter is made to follow a locus around the target volume as the gantry angle gradually changes to keep the beam tangential to the target. There is also a growing interest in the use of moving electron beams for irradiating the chest wall.[25]

At the RFH, some preliminary work has been carried out by Hu et al.[26] using the RFH tracking cobalt unit. Multiple wedged beams are directed tangentially toward the chest wall; a variant of this plan employs continuous irradiation as the isocenter travels around the chest wall and the gantry angle changes.

Treatment planning for such target volumes may be carried out using conventional planning systems provided that a "balancing" procedure is used to determine the dose weightings needed for each beam or at each gantry angle. It should be noted, however, that dose calculations for glancing fields may be significantly in error unless various corrections are made for the lack of scattering material at the outer edge of the tangential beam.

Target volumes curved in two planes may be irradiated by abutting or overlapping a series of single-plane plans having different curvatures in the transverse plane, as described previously in type C and type D treatments.

Figure 24. Examples of calculated dose distributions for three cross sections of a birfurcated pelvic target volume (redrawn and simplified from a Rad-8 printout). This plan indicates that some adjustment is required, since the isodose pattern is slightly too anterior, and the combined dose distribution for the modified plan is similar but complicated by having more beams and rotation centers. (a) Dose distribution at CT scan position 236 mm. (b) Dose distribution at CT scan position 268 mm. (c) Dose distribution at CT scan position 316 mm.

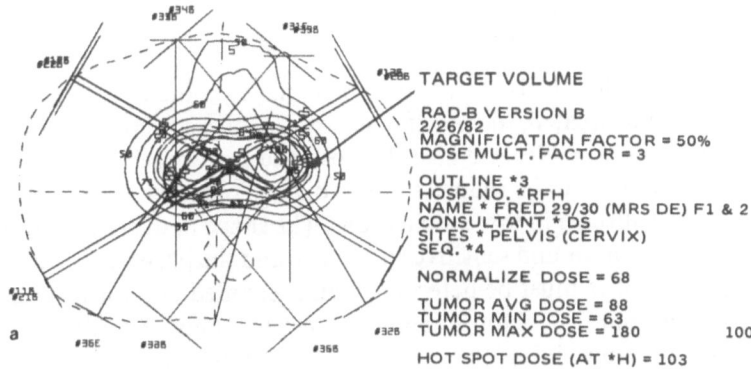

TARGET VOLUME

RAD-B VERSION B
2/26/82
MAGNIFICATION FACTOR = 50%
DOSE MULT. FACTOR = 3

OUTLINE *3
HOSP. NO. *RFH
NAME * FRED 29/30 (MRS DE) F1 & 2
CONSULTANT * DS
SITES * PELVIS (CERVIX)
SEQ. *4

NORMALIZE DOSE = 68

TUMOR AVG DOSE = 88
TUMOR MIN DOSE = 63
TUMOR MAX DOSE = 180   100

HOT SPOT DOSE (AT *H) = 103

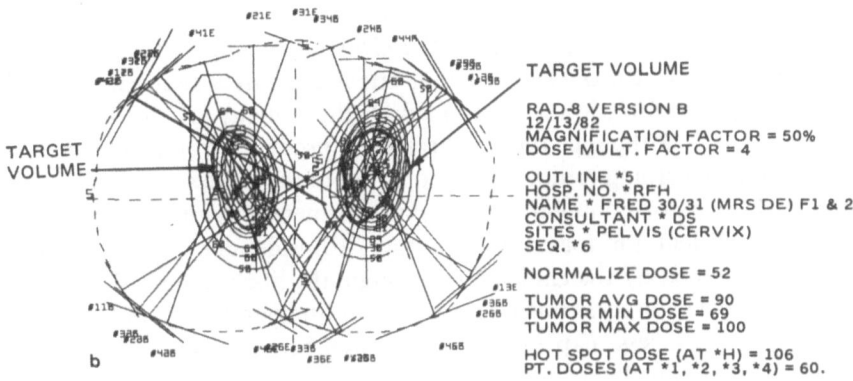

TARGET
VOLUME

TARGET VOLUME

RAD-8 VERSION B
12/13/82
MAGNIFICATION FACTOR = 50%
DOSE MULT. FACTOR = 4

OUTLINE *5
HOSP. NO. *RFH
NAME * FRED 30/31 (MRS DE) F1 & 2
CONSULTANT * DS
SITES * PELVIS (CERVIX)
SEQ. *6

NORMALIZE DOSE = 52

TUMOR AVG DOSE = 90
TUMOR MIN DOSE = 69
TUMOR MAX DOSE = 100

HOT SPOT DOSE (AT *H) = 106
PT. DOSES (AT *1, *2, *3, *4) = 60.

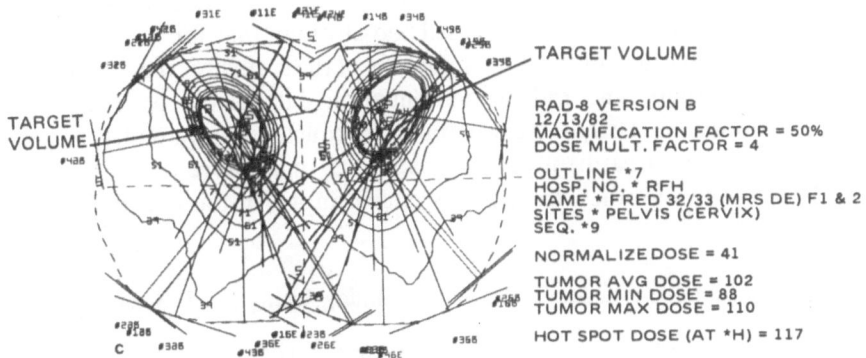

TARGET
VOLUME

TARGET VOLUME

RAD-8 VERSION B
12/13/82
MAGNIFICATION FACTOR = 50%
DOSE MULT. FACTOR = 4

OUTLINE *7
HOSP. NO. * RFH
NAME * FRED 32/33 (MRS DE) F1 & 2
SITES * PELVIS (CERVIX)
SEQ. *9

NORMALIZE DOSE = 41

TUMOR AVG DOSE = 102
TUMOR MIN DOSE = 88
TUMOR MAX DOSE = 110

HOT SPOT DOSE (AT *H) = 117

8. TUMOR (TARGET) LOCALIZATION

8.1. General Considerations

In order to find the best treatment plan, the radiotherapist must indicate the volume to be treated, the position of any sensitive structures that must be spared as far as possible, and also the dose and fractionation scheme. If the high-dose volume is to be shaped in three dimensions, then the patient's data must also be collected in three dimensions. The shape of the body outline, the position of the tumor and sensitive structures, and the position and density of any heterogeneities must be determined and recorded in a form suitable for three-dimensional calculation. Some allowance must be made for uncertainties in determining the position of the target volume and various internal structures, since patients may change in weight and tumors may shrink during the course of treatment.

The CT scanner is the most suitable device for tumor localization and measuring tissue densities, and since the CT scanner uses X rays, albeit of diagnostic quality, it provides radiation attenuation data that can be used with suitable modification at the much higher energy of the treatment machines.[27-29] Although the CT scanner produces three-dimensional information concerning radiological properties of the tissues and a reference framework for repositioning the patient for treatment, this information on its own is insufficient for locating the tumor. Considerable experience and expertise is needed by the clinician to estimate the likely extent of a tumor, and the diagnostic process must be regarded as an integral part of treatment planning. Nonetheless, experience from a number of centers shows that treatment plans are likely to be significantly altered when CT scans are obtained after conventional treatment planning.[30, 31]

Regrettably, the CT-scanner apertures are often too small to accept the patient in the treatment position, particularly if the upper half of the body is to be treated. It cannot be emphasised too strongly that for radiotherapy-planning purposes, one of the main functions of the scanner is measuring and recording the shape and radiation-attenuation data of the patient in the treatment position. Unsuitable examination couches compound this problem; couches for treatment-planning purposes must have a flat top and should not sag more than the therapy couch on which the patient is to be treated.

8.2. Tumor Localization for Treatment Planning Using a CT Scanner

In order to relate measurements on the CT scanner to the treatment machine control data, three reference planes must be defined within the patient that can be located relative to the CT scanner, the treatment-planning computer, and the treatment machine (Figure 25).

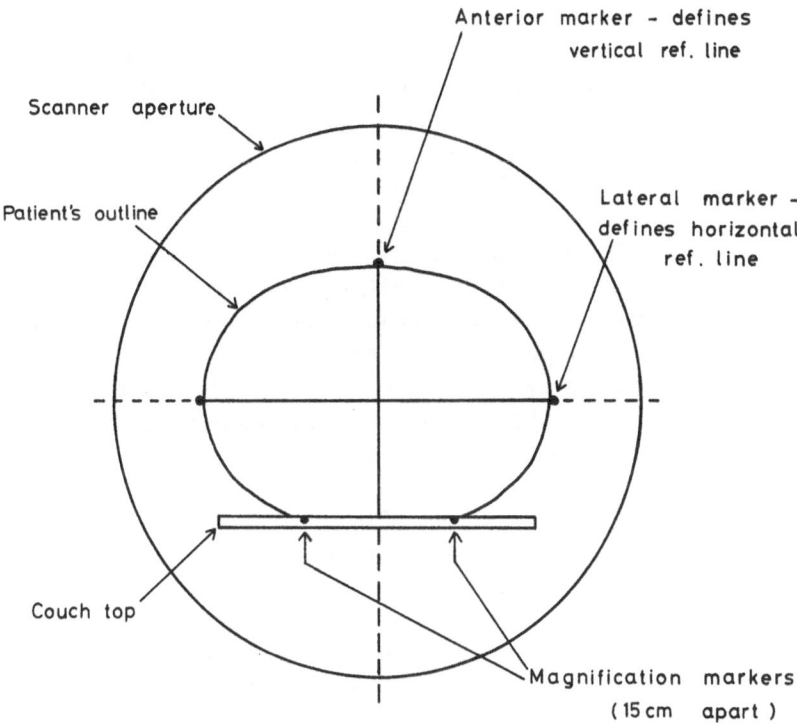

Figure 25. Diagram showing positions of the reference planes, CT scan markers, and magnification gauges used for treatment-planning purposes at the Royal Free Hospital, London.

With the patient lying in the treatment position on the scanner couch, three parallel lines are drawn on the skin with the aid of the scanner centering lights or using auxiliary wall and ceiling lights. One of these lines runs down the patient's mid line on the uppermost surface to indicate the anatomical median plane; it may be helpful to check the alignment of this marker by performing an AP scan with a radio-opaque strip along the (provisional) line. The other two lines are drawn at a fixed height above the couch top, one on either side of the patient, and they serve to define the horizontal (coronal) plane. The line of intersection of the median plane and the coronal plane will normally be coincident with the rotation axis of the untilted scanner. For treatment, the patient will be placed on the treatment couch, which is then positioned to bring the three reference lines into the desired relationship with the treatment unit isocenter and gantry rotation axis. For planning and treatment purposes, all lateral and vertical couch displacements will be expressed relative to these reference planes. A transverse reference plane is also selected and marked at the CT scanning stage, and all couch axial displacements are expressed relative to this plane.

Radio-opaque markers are taped along the reference lines so that they can be identified on the CT images; some radiographers also use a cross marker to show the position of the transverse reference plane, while others prefer a visual check of the scanner longitudinal scale when the marked reference plane is in the scan plane.

In addition to serving as reference lines from which the treatment unit isocenter displacements may be specified, the three axial reference lines also help the radiographer place the patient in the same position (posture) on the treatment couch each day. At the RFH, two long parallel markers 15 cm apart set into a card are placed under the patient to serve as magnification gauges and indicate the position of the couch top.

In some scanning systems, marked areas may appear as only an "overlay" that can be neither stored nor transferred to the planning system with the image data; in such cases, regions of interest will have to be entered directly into the planning system. Sometimes two or more target volumes will occur in the same CT slice, and each of them will have to be transferred to, and be identified by, the planning system, which should be able to perform the necessary dose calculations on each individual volume. Some therapists prefer to mark the target volume(s) and other regions of interest on film or other hard copy produced by the scanner; this information is then transferred to the planning system via a digitizer. This method precludes the use of the CT scan data for "pixel-by-pixel" calculations, and heterogeneities must be assigned a "bulk density" by the planner.

Once the necessary patient data has been stored in the planning system and the target volume and other areas identified, the planner prepares a suitable treatment plan in accordance with the prescription. Once the plan is accepted by the clinician, it is translated by system software into a data file for use by the computer that controls the treatment unit. This step is a vitally important link in the chain of events leading from the collection of scan data to the production of a safe and practical treatment file. It is of the utmost importance to ensure that the treatment unit is instructed only to carry out operations within its capabilities and that do not endanger the patient. The editor program will need information about the treatment machine geometry and its operating characteristics, such as gantry acceleration and deceleration, speed of collimator adjustment, dose rate, and so on. The absence of sophisticated software of this kind precludes using automatic treatment set-up and dynamic treatment on many commercial units having computerized checking and recording systems.

8.3. Treatment Simulation

Clearly, it is impossible to simulate conformation therapy treatments on conventional simulators. The only practical long-term approach to this problem involves using the treatment-planning system to carry out a computer simulation where patient data, machine geometry and control characteristics, and beam data can be combined to display the treatment process.

At present, considerable reliance is placed on phantom measurements using film, thermo luminescence dosimeters and other dosimeters to test the treatment plan and confirm that the treatment plan has been correctly transferred to the treatment unit control computer. These procedures are described in the following section.

9. EXPERIMENTAL VERIFICATION OF TREATMENT PLANS

9.1. Objectives

It is important to determine the objectives of any tests that are made. Such tests may be made to check the suitability of the plan to be used, confirm the accuracy of beam models and dose-calculation programs, verify that the treatment plan has been correctly transferred from the planning system to the machine control computer, or test the operation of the treatment machine.

Differences between calculated and measured dose distributions may be due to a number of causes.

1. There may be a difference between the plan used by the planning computer and the machine control computer, due perhaps to simple error, such as different plans being compared, or to the interpretation of the plan by the two systems. A simple example occurs regularly in conventional treatments when the treatment unit carries out a continuous-rotation treatment, but the planning computer computes doses for a number of fixed-beam stations. The effects of such differences in interpretation should be carefully considered for complex treatment plans.

2. The dose-calculation algorithms and beam models may be unsatisfactory. Problems may arise when doses are to be calculated at points outside the useful beam, as conventionally defined. Doses may be arbitrarily set at zero outside certain geometric limits, or there may be a failure to correct for changes in the beam profile with distance from the principal planes of the beam, with changes in the beam length or with tissue thickness. Errors arise in rotation treatments because of the limited number of stations selected and the limited beam "table" width. These errors may be significant when the beam is not directed through the same isocentric point throughout the treatment. Corrections for inhomogeneities may give rise to significant errors, as in conventional treatments.[32]

3. Even if the treatment plan has been correctly transferred to the treatment unit's control computer, an incorrect treatment may be given if the machine is incorrectly calibrated in terms of dose rate, beam sizes, and machine positions and speeds. The machine's control computer can only read data sent to it by the machine transducers—it has no way of knowing if they have been incorrectly set. If, for example, a collimator potentiometer has slipped on its shaft, an incorrect beam size will result.

No matter how many tests may be performed, it is well to remember that the basic aim of the treatment file is only to produce a specific radiation flux pattern in space relative to the machine isocenter. The patient must be positioned correctly relative to the isocenter by the radiographers. If the patient is placed prone instead of supine or with head instead of feet toward the gantry, the unfortunate results can be imagined. Further, the treatment unit, when in calibration, will deliver exactly the same treatment each day— even if it is wrong! It is essential for technicians to have a complete understanding of the treatment unit and the treatment plan. A "radiation-free" dummy run should always be performed before treatment begins.

9.2. Choice of Test Phantoms and Dosimeters

As stated previously, the only practical approach to routine treatment simulation for dynamic treatments is likely to be complete computer simulation. However, it will always be necessary to test new treatment techniques, and in the short-term phantom measurements are necessary to allow treatments to proceed. There are two basic approaches in using phantom measurements; viz.,

1. The dose distribution in the patient may be found by measuring the distribution in an anthropomorphic phantom, such as the Alderson Rando phantom[2], which has a processed real skeleton and artificial lungs. Such phantoms are never exactly like the patient, and a decision must be made about the equivalent place in the phantom for the target volume and the reference planes. The best solution usually is to place the target in the same position relative to the spine for patient and phantom.

Doses measured in the phantom must be converted to those that would be received by the patient. Strictly, a point-by-point comparison throughout the volume of interest is needed. Simple corrections may be made by using the phantom TAR/patient TAR ratios along the tumor axis—such corrections may be estimated using a conventional planning computer supplied with the dominant beam configurations for the phantom and patient outlines. This method also provides a check of the phantom and patient TARs over the target cross sections.

2. An alternative approach is to use a simple homogeneous phantom of

uniform outline and unit density to confirm that the computed and measured dose distributions are in agreement. Provided that the computer algorithms and calculation methods are carefully verified for use under these simple conditions, the phantom may be used to confirm that the treatment unit has performed as intended. This does not prove that the dose distribution in the patient will be as intended unless it has also been established that the planning system is able to calculate the three-dimensional dose distribution for a real patient (or phantom) with lungs and bones.

If a treatment plan is applied to both the Alderson or similar test phantom and to a unit-density phantom by the planning system and if both these phantoms are then irradiated using the treatment machine's interpretation of the plan, valuable information may be obtained. Once it has been proved that the planning system gives correct dose distributions for both the Alderson phantom and the unit-density phantom, it will be necessary only to use the unit-density phantom for routine tests and plan verification. If the treatment plan is applied to a standard phantom outline stored in the planning system and the same plan applied to a similar instrumented phantom by the treatment unit, the distributions may be directly compared.

9.3. Current Measurement Procedures

An Alderson Rando phantom is used at present to measure three-dimensional dose distributions; simple corrections are applied to modify these measurements to distributions in the patient. For a complete test of a dynamic treatment program, the whole treatment sequence should be executed exactly as for treatment. For film dosimetry, it is advisable to restrict the dose in order to maintain a reasonably linear dose-density response curve. For cobalt units, it is not practical to modify the treatment file so that it delivers a proportion of the actual daily treatment dose; the accuracy of speed control for couch and gantry, of collimator settings, and of the synchronization of machine movements depend on the dose to be given. Beam attenuators may be used to reduce the dose provided that they have no significant effect on relative doses over the target volume. Measurements must be made to determine any corrections to be made for beam hardening or changes in the beam profile. The relative effect of collimator leakage through and past the attenuator should be tested. On linear accelerators, it may be possible to deliver a balanced proportion of the dose over the same time by changing the pulse rate.

For routine tests, TLD LiF microrods 6 mm in length and 1 mm in diam. are inserted into small holes drilled in a matrix in each phantom slice. Ilford line film in thin plastic bags is placed between each slice along the target volume. Film has good spatial resolution but an unreliable dose-density response curve, dependent on processing conditions and emulsion batch. On the other hand, TLD rods can only be placed at relatively few positions over each slice, but they have a linear dose response and with proper calibration procedures give reproducible results with small standard deviations. Unfortunately, although it is possible to find out how much dose each microrod receives, it is not certain from this information where the microrod was situated with respect to the high-dose volume. While the intended pattern of dose distribution is known, it cannot be assumed that it has been achieved. Errors in positioning the TLD rods and the test phantom relative to the treatment unit can give a false impression of the effect of the treatment plan. The procedure adopted uses TLD rods to calibrate the films. Each film lies adjacent to an array of microrods. Situating pins on each phantom slice make pressure marks on the films that can then be precisely located with respect to the TLD rods. A relationship can then be established between the film density and the TLD reading for each film. It is assumed that for each film, points of equal density have received an equal dose.

Both the film and the TLD dosimeters are calibrated to a depth of approximately 10 cm in tissue-equivalent material. The same beam attenuator is used to irradiate the Alderson phantom and the calibration phantom. A complete calibration curve is usually obtained from the films.

9.4. Dose Measurements Using a Simple Phantom

In the tracking technique, the dose delivered is determined by the beam length and the tracking speed, which must be accurately controlled. It may be argued that with the exception of the axial dose profile, all aspects of the machine operation can be checked by direct observation. A simple procedure has therefore been devised to test both the treatment plan and machine operation by measuring the axial dose profile in a simple rectangular phantom. Although this procedure does not test all the machine parameters, it does give a rapid method for testing critical aspects of the treatment file, and it is used for carrying out reproducibility studies on the axial dose-controlling parameters and the effects of beam attenuators.

A copy of the patient's treatment file is modified so that all the parameters are fixed except those that control the axial dose distribution, viz., the collimator length, tracking speed and couch position, and the radiation shutter. The fixed parameters are set to provide a convenient geometry for irradiating the phantom. Typically, the gantry is set to direct the beam vertically downward, and the couch height and lateral position are adjusted to place a line of radiation detectors at the isocentre and parallel to the gantry rotation axis.

The test phantom is designed to accommodate a high-dose volume 40 cm long and 14 cm wide with a 5-cm margin all around to provide scatter. The phantom is made up of 2.0-cm-thick slabs of Perspex, one of which has been drilled with holes at 1-cm intervals to accommodate "Therados" DPD-5 semiconductor probes. An additional 1-cm slab is also available, and the thickness of material overlying the line of detectors can be adjusted in 1-cm steps from 1 to 20 cm. For good spatial resolution and a permanent record, the phantom will accept film in plastic bags, with each phantom section provided with locating pins that also pressure mark the films. One phantom slab has a shallow milled groove along the midline to take pairs of TLD microrods.

REFERENCES

1. J. B. Massey, Radiotherapy treatment planning, *Post. Fiz. Med.* **14**, 33–46 (1979).
2. S. W. Alderson, L. H. Lanzl, M. Rollins, and J. Spira, An instrumented phantom system for analogue computation of treatment plans, *Am. J. Roentgenol.* **87**, 185–195 (1962).
3. S. Walbom-Jorgensen, L. Cleeman, A. Nyb-Rasmussen, and P. Byrge Sørensen, Technical aids for radiotherapy of Hodgkin's disease, *Br. J. Radiol.*, **45**, 949–953 (1972).
4. T. Mori, Conformation radiotherapy with a NELAC-1006 (Clinac-6), in *Proceedings of the Eighth Varian Clinac Users Meeting Jan. 31–Feb. 2, 1980, Kauai, Hawaii*, Varian Associates, Palo Alto, 1980, pp. 43–44.
5. T. Matsuda, Computer-controlled conformation therapy, in *Proceedings of Medinfo 80 Japan*, D. A. B. Lindberg and S. Kaihara eds., North-Holland, Amsterdam, 1980, pp. 14–17.
6. H. Perry, J. Mantel, and J. J. Weinkam, Optimized treatment planning: automatic control of some parameters, in *International Symposium on Fundamentals in Technical Progress, Liege, Belgium, May 1979*, vol. 2, Presses Universitaires de Liege, pp. 13.1–13.35 (1979).
7. B. S. Proimos. Synchronous field shaping in rotational megavoltage therapy, *Radiology* **74**, 753–757 (1960).
8. B. S. Proimos, New accessories for precise teletherapy with cobalt-60 units, *Radiology* **81**, 307–316 (1963).
9. S. Takahashi, Conformation radiotherapy: rotation techniques as applied to radiography and radiotherapy of cancer, *Acta Radiol. Suppl.* 242, 1–42 (1965).
10. J. G. Trump, K. A. Wright, M. I. Smedal, and F. A. Saltzman, Synchronous field shaping and protection in 2-million-volt rotational therapy, *Radiology* **76**, 275 (1961).
11. J. A. Brace, A computer system for the dynamic control of a tele-cobalt unit, in *International Symposium on Fundamentals in Technical Progress, Liege, Belgium, May 1979*, vol. 2, Presses Universitaires de Liege, pp. 12.1–12.18 (1979).
12. J. A. Brace, T. J. Davy, and D. B. L. Skeggs, A computer-controlled cobalt unit for radiotherapy, *Med. Biol. Eng. Computing* **19**, 612–616 (1981).
13. J. A. Brace, A computer-controlled tele-cobalt unit, *Int. J. Radiat. Oncol. Biol. Phys.* **8**, 2011–2013 (1982).
14. A. Chung-Bin and T. Wachtor, Applications of computers in dynamic control of linear accelerators, in *Proceedings of the Third International Computer Software and Applications Conference 1979*, IEEE, New York 1979, pp. 270–275.
15. B. Bjärngard, P. Kijewski, and C. Pashby, Description of a computer-controlled machine, *Int. J. Radiat. Oncol. Biol. Phys. (Suppl. 2)* **2**, 142 (1977).
16. M. B. Levene, P. K. Kijewski, L. M. Chin, B. E. Bjärngard, and S. Hellman, Computer-controlled radiation therapy, *Radiology* **129**, 769–775 (1978).

17. L. M. Chin, P. K. Kijewski, G. K. Svensson, J. T. Chaffey, M. B. Levene, and B. E. Bjärngard, A computer-controlled radiation therapy machine for pelvic and para-aortic nodal areas, *Int. J. Radiat. Oncol. Biol. Phys.* **7**, 61–70 (1981).

18. J. van de Geijn, A computer program for three-dimensional planning in external-beam radiation therapy, EXTDØSE, *Comput. Programs Biomed.* **1**, 45–57 (1970).

19. J. R. Williams, Dosimetric problems in moving-table cobalt-60 teletherapy, M.Sc. thesis, University of London, London, England (1973).

20. T. J. Davy, P. H. Johnson, R. Redford, and J. R. Williams, Conformation therapy using the tracking cobalt unit, *Br. J. Radiol.* **48**, 122–130 (1975).

21. J. van de Geijn, EXTDØS 71, revised and expanded version of EXTDØS, A program for treatment planning in external-beam radiotherapy, *Comput. Programs. Biomed.* **2**, 169–171 (1972).

22. A. Green, W. A. Jennings, and H. M. Christie, Radiotherapy by tracking the spread of the disease, in *Transactions of Ninth International Congress of Radiology, Munich 1959*, Urban and Schwarzenberg, Munich–Berlin, 1960, 766–772.

23. B. S. Proimos, S. P. Tsialas, and S. C. Coutroubas, Gravity-orientated filters in arc cobalt therapy, *Radiology* **87**, 933–937 (1966).

24. A. Chung-Bin, T. Wachtor, T. Zusag, and F. Hendrickson, The treatment of mesothelioma of the pleura with computer-controlled dynamic treatment, in *Proceedings of the Fifth International Conference of Medical Physics and Twelfth International Conference of Medical and Biological Engineering, 1979*, vol. 1, Bellinson Medical Center, Petak Tikva, Israel, 1979, p. 1600.

25. L. H. Lanzl and M. Rozenfeld, Automatic three-dimensional dosimetry in electron beams, *International Congress Series No. 339, Radiology*, vol. 2; *Proceedings of the Thirteenth International Congress of Radiology, Madrid, October 15–20, 1973*, Excerpta Medica, Amsterdam, 1974, pp. 650–652.

26. Y. M. Hu, T. J. Davy, and D. B. L. Skeggs, A multiple tangential field technique for the treatment of the chest wall using a computer-controlled tracking unit: a preliminary study (to be published).

27. R. P. Parker, P. A. Hobday, and K. J. Cassell, The direct use of CT numbers in radiotherapy dosage calculations for inhomogeneous media, *Phys. Med. Biol.* **24**, 802–809 (1979).

28. M. R. Sontag and J. R. Cunningham, Clinical application of a CT-based treatment-planning system, *Comput. Tomog.* **2**, 117–130 (1978).

29. P. K. Kijewski and B. E. Bjärngard, The use of computed tomography data for radiotherapy dose calculations, *Int. J. Radiat. Oncol. Biol. Phys.* **4**, 429–435 (1978).

30. M. Goitein, Benefits and cost of computerized tomography in radiation therapy, *J. Am. Med. Assoc.* **244**, 1347–1350 (1980).

31. J. E. Husband and S. J. Golding, Computed tomography of the body: When should it be used? *Br. Med. J.* **284**, 4–8 (1982).

32. J. R. Cunningham, Tissue inhomogeneity corrections in photon-beam treatment planning, in *Progress in Medical Radiation Physics*, vol. 1 (C. G. Orton, ed.), pp. 103–131, Plenum, New York (1982).

1-III

Computer Systems for the Control of Teletherapy Units

J. A. BRACE

1. THE TELETHERAPY UNIT

1.1. The Royal Free Hospital's Tracking Cobalt Unit

During the latter part of 1979, a computer-controlled tracking cobalt unit was installed at the Royal Free Hospital, London. It is based on a standard TEM MS90 unit and operates at 90-cm source-axis distance with a geometric field size of 45 × 45 cm at that distance. It has been modified so that it can be used either manually or under computer control. The first modification added drives for the couch lateral and couch longitudinal movements; the longitudinal drive may be operated in positional mode or rate mode. The second modification added the computer interface and servo mechanisms as shown in Figure 1. The final modification was the addition of a small side panel to the radiographers control console, which contains an on/off line switch and a key switch that must be set in order for the shutter to be operated by the computer.[1-3]

There are nine parameters that can be controlled positionally and two that can be controlled in rate mode; they are given in Table 1. The position of all the moveable parameters can be interrogated by the computer, but the speeds can be verified only by timing the movement.

J. A. BRACE • Radiotherapy Department, Royal Free Hospital, Pond Street, London NW3 2QG England.

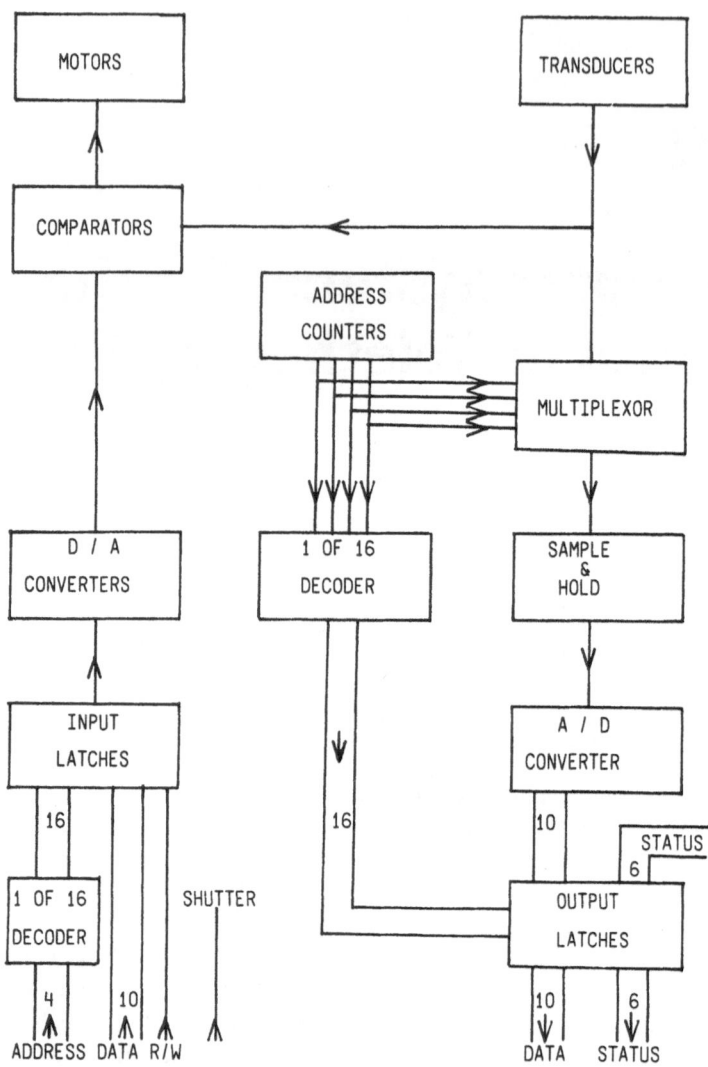

Figure 1. A simplified diagram of the computer-controlled cobalt unit.

1.2. Other Computer-Controlled Teletherapy Units

Most other teletherapy units that have been adapted for computer control are linear accelerators.[4-8] Besides the positional parameters, energy and pulse rate can often be controlled as well. For a cobalt unit, speed is defined in terms of distance and time, whereas for a linear accelerator, the

Table 1. Parameters of the Cobalt Unit That Can Be Controlled by the Computer

Parameter	Range	Resolution
Collimator X	4.0–44.0 cm	0.1 cm
Collimator Y	4.0–44.0 cm	0.1 cm
Collimator rotation	−100.0–100.0 deg	0.4 deg
Gantry rotation	Continuous	0.4 deg
Couch top longitudinal	0.0–85.0 cm	0.1 cm
Couch top lateral	−20.0–20.0 cm	0.1 cm
Couch height	−51.0–1.0 cm	0.1 cm
Couch–floor rotation	−100.0–100.0 deg	0.4 deg
Head rotation	−177.0–177.0 deg	0.4 deg
Couch longitudinal speed	−20.0–20.0 cm/min	0.0784 cm/min
Gantry arm speed	−360.0–360.0 deg/min	1.4117 deg/min
Shutter	Open/closed	

analogous parameter is defined in terms of distance and number of pulses. This means that by varying the pulse rate, it is possible to limit the physical required speeds to within a reasonable and safe range. Independently controllable collimators are another refinement sometimes incorporated in computer-controlled linear accelerators.

The control requirements are basically very similar in all computer-controlled units; however, the methods by which these requirements are met differ greatly. A computer-controlled unit consists of three main components; firstly, motors, transducers, or tachometers; secondly, a servo mechanism to control the motors meet a specified demand; and thirdly, a system to generate the demands that will produce the required treatment. The major difference between control systems currently in use is the form of the second component, viz., the servo mechanism. The Royal Free Hospital's system uses hardwired electrical engineering techniques to form the servo system, whereas other systems use specially written programs and analogue-to-digital convertors to perform the servo's function.

Neither system is ideal. The electrical engineering solution uses tried and tested techniques, but it does involve a greater additional cost when the computer-controlled unit is bought. The programmers' solution is far less expensive initially, since the only additional hardware is usually a multiplexor and an A/D and a D/A convertor; however, the programming required is rather complex and usually involves empirically determined constants.[4]

The ideal system for conformation therapy would be one where the computer specified the position and state that the treatment unit was to acquire in the next x pulses or seconds and the control system ensured that all parameters moved linearly to this position during the specified interval.

2. INTERFACE WITH THE COMPUTER

2.1. The Royal Free Hospital's System

To the computer, the treatment unit is a standard 16-bit parallel duplex device: There are 16 data lines to the device and 16 from the device; a "flag" line to tell the device that the data is valid, and a "ready" line from the device to tell the computer that it has received the data and that the data on the 16 lines into the computer is valid. The 16-bit word from the computer consists of three sections, a shutter bit (bit 15), data (bits 14–5), and an address (bits 4–0). Each of the nine parameters has a read address and a write address. Outputting the write address causes the data in bits (14–5) to be latched into the servo demand memory; outputting the read address causes the latest reading of the addressed parameter to be placed on data lines into the computer. Certain parameters have a read address but no write address, e.g., the couch top rotation; other parameters have a write address but no read address, e.g., the gantry speed.

Data from the treatment unit consists of two sections, status bits (15–10) and data (9–0). Data is the value requested by the last read request, and the status bits specify shutter open, treatment start pressed, on-line, ready, head pin in, and shutter closed, respectively.

One write address specifies which drives are to be active (one bit for each of the nine drives), and this word is used to deactivate all the drives and close the shutter (bit 15 set at 0) should any error occur. This "activate drives" word also means that all demands can be set up and then all the drives activated simultaneously. Two other write addresses are used to switch the unit on and off line.

All computer-controlled equipment should have some fail-safe mechanism in case the computer or the link to the computer breaks down. The Royal Free Hospital's system uses a "twitcher" device whereby a special sequence of words must be sent to the therapy unit at least once a second. If this sequence is not received, then a latch drops out, deactivating all the drives and closing the shutter. The sequence of words is devised so that open or short-circuited lines from the computer to the treatment unit result in the correct twitcher sequence not being received, thus dropping the latch.

The method of data transfer is very reliable and safe. It is also a very standard form of computer-device interface. With hindsight, it could be improved; for example, the addition of a parity bit would reduce the need for complex data verification procedures. This method is also limited to a maximum of 16 controllable and readable parameters if a data width of 10 bits is to be retained.

2.2. Other Data Transfer Methods

If the servo mechanism is not included within the treatment unit, then signals between transducers, motor drives, etc., and the computer may be in analogue form, with the computer containing the necessary analogue to digital convertors or alternatively the treatment unit containing the A/D convertors so that digital data is transferred between the computer and the device.

Digital signals may be transferred in parallel form as just described or serially. Serial-data transfer involves only three lines in its simplest form, although seven wires are needed to meet RS 232C specifications, as opposed to 36 lines for a parallel interface. Serial transfer is slower than parallel, and special protocols are needed to ensure that data over run (i.e., the fastest device continues to transmit data even though the receiver is too busy to receive it) does not occur. These special protocols (ENQ, ACK, X-ON, X-OFF, etc.) mean that transferring pure binary data is complex and differs from computer to computer; however, even allowing for the extra circuit necessary to convert the serial bit stream into words, the savings in cable costs usually make serial the cheapest option.

The fourth method of transferring data is a standard interface known as IEEE 488, GP-IB or HP-IB. This interface was specially designed for instrument control and monitoring, and it is becoming very popular on nearly all computers. In essence, this 8-bit parallel interface has eight data lines, three control lines, and five interface management lines. Up to 15 devices can be connected to the bus, with each device designated as a talker or listener or both. Data lines contain either data or an address depending on the state of one of the interface management lines.

Data transfer is very rapid and accurate. If computer-controlled therapy units are to become common, then a standard interface must be used, and the GP-IB instrument interface is a prime candidate.

2.3. Data Format and Codes

Data within the computer is held as 8, 16, or more bit words. Data between a computer and a device may be sent directly in binary form, or alternatively, it may be encoded prior to transmission and then decoded when received. All codes involve some degree of redundancy, i.e., the number of bits necessary to transmit coded information is more than that required to hold the original information. This redundancy is used to form control codes, e.g., ENQ, ACK, etc. The most common code is ASCII, which uses 8 bits for each character; thus, a three-digit number is represented as three 8-bit characters, i.e., 24 bits, whereas only 10 bits are needed to store the original information. Other coding schemes are available, e.g., ECBDIC, BCD, etc.; however, they are less common.

Not only numerical values of parameters, but also a method by which the therapy unit can associate data with particular parameters must also be transmitted. This can be done in two ways: either all data for all parameters is transmitted every time, or some way is devised by which data (sent) is associated with the correct parameters. The former method is very time consuming and error prone in its simplest form, since the computer and therapy unit may not remain synchronized. The second method involves either preceding data with a special code or as with the Royal Free Hospital's system, including the data addresses within the code.

2.4. Noise Detection and Elimination

All therapy units work in electrically noisy environments. Although every effort is made to reduce noise to a minimum, methods must be incorporated to detect noise and if possible eliminate its effects. The Royal Free system has no other way of detecting that noisy data arrived at the treatment unit than by carefully monitoring the action of the unit. Noisy data from the unit is more easily detected and corrected. Each parameter is read up to eight times until three consecutive identical readings are obtained (using the parallel interface, this takes less than 20 µsec). This "noise-free" reading is then compared with the last reading to verify that it is reasonable. If it is not, then up to three more noise-free readings are taken. If a reasonable reading still cannot be obtained, it is assumed that a fault has occurred, and the therapy unit is halted.

3. PROGRAMMING THE THERAPY UNIT

3.1. Basic Methods

Two basic methods are available for describing the action required by the treatment unit. The simplest method describes the motion as the position at set intervals, for example, every 5 sec. The alternative method used at the Royal Free Hospital describes the motion as a series of states, including time, to be reached before the next state is demanded. Typically, these "control vectors" will specify the next position and the exposure time to be taken. If the shutter is closed, then the time is zero, otherwise, the control program will adjust speeds if necessary so that all parameters reach their specified position at the specified time. If the specified position is the same as the current position, then the unit stays where it is for the time specified. The position of the unit is, of course, verified continuously throughout treatment.

Each control vector specifies the shutter state; collimator sizes and rotation; the couch longitudinal, lateral height, and floor rotation; gantry position; head rotation; the couch longitudinal and gantry speed; and the time to be taken. These control vectors are created by the editor program and checked to be feasible.

3.2. Modifications to the Basic Method

Ideally, all drives would be speed controlled and infinitely variable and accurate; this not being the case, some modifications are necessary. Firstly, only two of our drives can be speed controlled; for the others, a pseudo speed routine is used, i.e., the required motion is divided into small steps, and the control program automatically issues the total movement as a series of small steps at the requisite times. Secondly, if the control vector implies couch longitudinal movement and gantry movement, then it is unlikely that both demands will be met at the same time. Typically, the couch speed is 30 mm/min, i.e., 1 mm in 2 sec; the arm speed is 120°/min, i.e., 1° in 0.5 sec; thus, the couch longitudinal will have the same reading while the arm travels 4°. For this reason, a "fit factor" is generated as a function of the distance the arm has to travel, the distance the couch has to travel, and the time left. The next control vector is activated when this fit factor reaches a minimum. Thirdly, speeds can be defined in units of only 0.784 mm/min or 1.41°/min, and this is often not accurate enough. Speeds are slightly adjusted throughout the step to ensure that the arm and couch longitudinals arrive at the specified positions when the time interval has expired.

The use of control vectors greatly reduces the data necessary to define a particular treatment; it does, however, imply that all positions are linearly interpolated between given positions, which is usually satisfactory. However, in order to geometrically cover an ellipse while rotating the gantry from minor to major axis, the collimator motion should not be a linear function of the gantry angle, but a complex function of the tangent of the angle; in practice, the linear interpolation between the major axis diameter and minor axis diameter is sufficiently accurate. However, if necessary, an intermediate control vector at 45° can be created.

3.3. Base Position and Safe Starting Positions

All couch positions, i.e., long, lateral, and height, are specified relative to a base position. When the patient is set up using skin marks, the set-up position is used as the reference so that all couch movements are relative to the set-up position. The longitudinal base position is arbitary; however, the lateral and height base positions define a safe range for the couch when the patient is set up. The safe range for the couch height is defined initially as within 1 cm of the height calculated, so that the patients' skin marks coincide with the machine isocenter. The editor program simulates movements involved throughout the treatment and from a knowledge of the geometry of the complete unit, the range in which the couch lateral can safely be calculated. The patient is assumed to be "contained" within a box of width equal to the couch width (28 cm) and of height 40 cm. In practice, collisions are more likely to occur between the couch and arm—especially if the wedge box is used. If a safe lateral range is impossible with a 1-cm tolerance on the height, then this tolerance is reduced to 5 mm and then to 2 mm before the treatment file is flagged as unsafe.

Since our unit has only 90 cm source–axis distance (SAD), the distance from the face of the collimators to the isocenter is only 45 cm, or 37 cm if the wedge box is attached; hence, collision avoidance is critical. The collision program works only if the couch top and the couch floor rotations are zero.

4. THE COMPUTER SYSTEM

4.1. Hardware

At the RFH, we use two separate computer systems—one to actually control the unit, located under the control desk; the other used by the physicist to produce the treatment files, i.e., the set of control instructions for the treatment unit. The two computers are linked by a 9600 baud RS 232C interface.

The computer system actually used to control the treatment unit is a Hewlett-Packard L-series micro with 128 Kbytes of memory, a 12-Mbyte Winchester, a small printer, and a terminal. The system is configured so that it automatically boots when the power is turned on, and a menu is displayed on the screen; thus, the radiographer or physicist can, by simply pressing a function key, execute the RUN-UP program, the actual CONTROL program, the DIAGNOSTIC program, the UTILITY program or the LINK program. All programs are pass-word protected.

The treatment-planning system is a Hewlett-Packard series 1000 model 45 with 128 Kbytes of memory, a fast floating point processor, 14.7 megabyte hard disk, calligraphic display, color rastor display, plotter, printer, digitizer, two terminals, and a tape unit. As for the microcomputer, the boot-up produces a menu, so that the physicist can, by pressing function keys, select the EDITOR program, the UTILITIES program, or the LINK program. When the required program has executed, the menu reappears for the next choice. This set of production programs uses only the printer, hard disk, and one terminal; all the other peripherals are used by the development system described later.

4.2. Software

4.2.1. FILES

Both systems at the Royal Free Hospital use treatment files. The planning system creates a file, which is transmitted to the control system prior to the treatment start date. When treatment has been completed, parts of the treatment file are transmitted back to the planning system. The fact that the treatment file may be stored in two different systems simultaneously necessitates using an ownership flag associated with each treatment file to prevent a file being modified in the planning system while in use in the control system.

The control system has a journal file that records all the activity on the control system; e.g., the treatment of a patient is recorded along with the time, the file number, the total exposure time, and whether the treatment ended normally or not. This journal can be listed and cleared by only the computer scientist using her or his special pass word.

During the course of treatment, a log record is written to the hard disk every 1 or 2 sec. This record contains sufficient information so that if a power failure occurs, treatment can be recommenced from the point of power failure. This log record is also used as an error log. If an error occurs, then the position and state of the unit is recorded along with an error number. Prior to the start of any treatment, the log record is read and treatment can be started only if the error number is zero; the error log can be cleared by only an authorized physicist or the computer scientist. There are over 100 separate error codes covering machine faults, operator errors, software errors, and computer hardware errors. Any drive failure causes a journal record to be written as well as a log record. The journal record specifies precisely which drive is faulty and what the symptoms are.

4.2.2. EDITOR PROGRAM

The editor program enables the physicist to produce a patient treatment file. Each patient may have up to ten treatment files, each of which consists of four types of records. Type 1 records contain prescription data; e.g., dose per treatment, dose to date, date of next treatment, etc. This record is initially created by the physicist using the editor program, but it is updated at the end of each treatment by the control program. Type 2 records contain set-up data, e.g., safe couch range, wedges that are needed, etc. This record is created by the physicist and the collision protection program. The type 3 record is the restart record. This record is normally null, but if treatment has been stopped before normal completion for any reason, e.g., the patient needs immediate attention, then the control program automatically stores the position and state of the unit so that treatment can be continued if required. There is only one of each of these record types per treatment file; however, there may be up to 32,000 type 4 records, which are the control vectors described previously. Each vector must be created by the physicist; however, certain aids are available. For example, the program assumes that the next position is the same as the current position, therefore the physicist only has to enter the changed values; the program will calculate the arm speed for a given movement and time, allowing for the finite acceleration and deceleration of the arm. A complete treatment file for a simple three-field treatment can be produced from the plan in about 20 min.

The editor program allows treatment files to be created, edited, copied, flagged as not to be used, listed, purged, or modified (e.g., couch height increased by 1 cm throughout or dose increased by 10%). If a treatment file has been sent to the control system or used, then it cannot be changed or purged, although it can be copied to another file.

4.2.3. PLANNING-SYSTEM UTILITY PROGRAM

The planning-system utility program enables the physicist to list all the patients and determine which files exist and which system has them (i.e., planning system or control system). This program also enables treatment files to be archived and restored by using tapes.

4.2.4. LINK PROGRAMS

The two link programs, one on the control system and one on the planning system, enable treatment files to be transferred between systems. During the transfer of treatment files, the control system microcomputer is slaved to the planning system minicomputer. The physicist on the planning system activates the link program by pressing a function key; the radiographer then activates the control system link program by pressing a function key. The control system terminal is now disabled until the link is terminated by the planning-system computer. The physicist on the planning system can transmit treatment files to the control system and receive the type 1 record (the updated dose record) and the type 3 record (the restart record) from the control system. Treatment files are purged from the control system when they have been successfully received by the planning system.

4.2.5. RUN-UP PROGRAM

The run-up program is a control system program that must be run every morning prior to the start of any computer-controlled treatment. This program is activated by the radiographer pressing the function key, and it operates all the drives in positive and negative directions and in the case of the arm and couch longitudinals, in both rate and positional modes. In this way, all the drives are thoroughly checked prior to any treatment. Similarly, by opening and closing the shutter five times in rapid succession, the shutter mechanism is tested. The run-up program includes a 2-min period during which all the values of all the variables are read as quickly as possible (over 1000 times per sec) to detect any noise on the lines. Should any error occur, an error log is written, which prevents the unit from being used until the error has been investigated.

4.2.6. CONTROL PROGRAM

The control program is primarily intended for use by radiographers, and it is deliberately designed to follow normal working procedures as closely as possible. It is essential to have a test mode of operation, so that the patient's treatment file can be run through without updating the patient's treatment record. This special test mode is entered by prefixing the patient's name with two asterisks; prescription details are not checked or updated. In test mode, the system also includes facilities for simulating only the shutter opening, zeroing all displacements, and logging actual end-point positions. Having entered the patient's name at the terminal and verified the identification, the radiographer enters the treatment file number. The program checks to see that this treatment is valid, i.e., due today, and that the prescribed dose will not be exceeded. If some inconsistency is discovered, e.g., the patient should have been treated yesterday, then a warning message is displayed. If the inconsistency is due to only dates, probably caused by public holidays, then the radiographer can override the warning and continue treatment. If, on the other hand, this treatment would cause the prescribed dose to be exceeded, then the radiographer cannot override the warning.

The computer next displays the set-up data on the screen, which is relayed to a monitor in the treatment room. These data are taken from the type 2 record and include patient-positioning information, wedge numbers, and the safe range for the couch. The computer reads the position and the state of the treatment unit continuously, waiting for the "ready" line to rise. The computer continually updates the display, and if any parameters are incorrect (e.g., wrong wedge) or the couch is in an unsafe starting position, this particular parameter is highlighted in inverse video on the monitor.

The patient is placed on the couch in the usual manner, the wedges are interlocked, the timer is set, the door interlocked, and the shutter key switch turned. Thus, all the normal safety features of the unit still apply. When the interlocks, etc., have been set, the computer detects the raised ready line and has a final check on the parameters. If any of them are found to be incorrect, then the radiographer can choose either to set up again or abort the treatment; only under the test mode can the "treatment" continue. The radiographer is now instructed to switch over the treatment unit to computer control—alternatively, to unready the unit if treatment is to be aborted. Once the program detects that it is in control, it sends a "program active" command to the unit, which from now on must ensure that the twitcher is activated at least once a second. The radiographer now presses *treatment start*. Under computer control, *treatment start* does not open the shutter; instead, the treatment start line rises. The computer now moves the unit to begin treatment position and gives the prescribed treatment. Throughout the treatment, the position and state of the unit are displayed on the terminal screen. At the end of the treatment, the treatment record is updated, and the radiographer is asked to press *treatment stop*, then switch to manual control, and then unready the treatment unit.

At any time during the course of treatment, *treatment stop* may be pressed. This button, unlike *treatment start*, does directly close the shutter as well as signal the computer. Thus, if the patient needs immediate attention, the radiographer can quickly press *treatment stop* and enter the room. Having detected *treatment stop*, the computer displays a message and waits for either *treatment start* to be pressed, in which case treatment continues as if no interruption has occurred, or alternatively, for the ready line to drop, which, of course, occurs if the radiographer breaks the interlocks by entering the room. Having attended to the patient, the radiographer can either cancel the treatment for the day, or set up the patient again in the original position and instruct the computer to complete the treatment. In the latter case, the computer will move the unit directly to the point of interruption and resume treatment while making allowances for the probable change in the reference position. If treatment has been canceled for the day, it may be completed next time. During treatment, the log record is written at frequent intervals, and the journal is updated with treatment details at the end of the treatment.

4.2.7. DIAGNOSTIC PROGRAM

There are, in fact, ten different diagnostic programs ranging from a very low-level program that enables individual words to be created, transmitted, and received, to a program that runs the unit for 1.5 hr and determines the constant of motion (e.g., acceleration constant, time to reach maximum speed, maximum speed, etc.) for all drives. The shutter timing can also be tested over long periods of time if required.

The most often used diagnostic simply enables the operator to create a control vector, i.e., specify position and speeds and then display the position of the unit as it moves to the requested position. One of the diagnostics used routinely de-activates the twitcher during movement in order to verify that the safety mechanism is operating correctly. The existence of such a comprehensive set of diagnostic tools and the error log and journal greatly contribute to the very little down time we experience.

4.2.8. CONTROL SYSTEM UTILITY PROGRAM

The utility program can be operated on either a low-level security to list the journal, display or print treatment files, examine the log record, etc., or it can be used by a higher security operator actually to clear the journal or the error log. Clearing an error that occurred during treatment causes a restart record (type 3) to be created so that the patient's treatment can be continued to normal completion if required.

5. SAFETY AND CONTROL MONITORING

5.1. Introduction

Using a computer to control a teletherapy treatment unit introduces additional safety considerations. Errors and faults may occur in four areas: human error by the radiographer or physicist, computer error in either hardware or software, malfunctions of the treatment unit, and finally faults in the links between the treatment unit and computers. At the Royal Free Hospital we have studied each possible source of error and tried to reduce its likelihood of occurrence as much as possible.

Errors in the treatment unit can be due to either complete failure of a drive or, more likely, to the drive continuing to operate while out of tolerance as either a rate control drive or a positional drive. The tolerances used in the programs depend on the particular action; e.g., if the shutter is closed, i.e., between fields, a tolerance of 0 mm and 0° is demanded. If after 10 sec a position has not been reached, a tolerance is expanded to 1 mm and 0.4°; if this tolerance cannot be satisfied after 30 sec, then an error is flagged. During actual exposure, the tolerance allowed varies with the parameter but is typically 1 mm and 0.8°.

5.2. Computer Hardware Errors

There are two computers in the system at the Royal Free Hospital both of which are capable of malfunctioning. In practice, it is very unlikely for the actual processing unit and memory to malfunction other than to stop working altogether. Most modern computers rigorously check themselves every cycle for memory failures and can be set to go into a "hard halt" on failure. Therefore, if the control system computer should fail during treatment, it will halt completely; the twitcher will stop, causing the treatment unit to halt and close the shutter.

The most common hardware faults are associated with mechanical devices, such as printers, disc drives, etc. The control program operating in normal mode does not use the printer, only the terminal and Winchester disk drive. The Winchester disk has proved to be very reliable, but as an extra precaution, all records, each 32 words long, contain a "checksum" word, which is generated and checked every time the record is read.

5.3. Computer Software Errors

The main programs at the Royal Free Hospital have now been in use for over three years, and nearly 10,000 fields have been given without any software errors being detected, but no doubt there are still some bugs in the code yet to be discovered, particularly in the less critical parts of coding.

The treatment file is produced by the editor; the control program assumes that it is correct and apart from checking the checksum word when reading the file, the control program does not check the file. This means that the control program is entirely devoted actually to controlling the unit and interacting with the radiographer. The editor, on the other hand, has to know the characteristics of the treatment unit, speeds, etc., in order to check the feasibility of the control vector sequence and also to calculate the time necessary for each requested movement. The editor program, therefore, is very critical, and it must be very reliable and error free. This has been achieved by highly modular programming, structured as much as possible with FORTRAN. A secondary program is used to check the primary program, which interacts with the planning physicist to produce a treatment file, checked for validity and feasibility as it is being created. Prior to use, the treatment file must also be verified by the secondary program, which has no subroutines in common with the primary program. It is hoped that most software errors are detected by this method.

5.4. Treatment Unit Errors

The state and position of the treatment unit is checked every 1 sec or less; originally, tests were based on estimating the distance the parameter should have traveled since the beginning movement. This proved not to be very satisfactory, since calculations of the estimated position took a very long time and also the necessary tolerance on the fast moving parameters, e.g., the collimators, could become greater than the expected movement. Most drive faults either cause the drive to stop or not to reach the specified position in time. Tests have therefore been modified to detect no movement when there should be movement and movement when there should be none. This information is backed up by verifying that the position is reached within the specified tolerance at the calculated time. Typical faults that have been detected and handled safely include sticking collimator blades, faulty wedge interlock switches, and excessive oscillation or "hunting" of the floor rotation.

When any error in either hardware or software is detected, the "fast-path" return sequence part of the control program ensures that the error-handling routine is executed immediately. The error handler first issues a "stop drive close shutters" command and then continues to monitor the treatment unit. The terminal displays a message to the effect that the computer is stopping treatment. If the computer detects that the shutter will not close, it stops sending the twitcher, thereby invoking the treatment unit's own close-down procedure, and also sends the "program inactive" word as another way of invoking the close-down sequence. Since all normal safety mechanisms are operative, the emergency stop buttons could be used by the radiographer as a last resort. Having safely halted the treatment unit, the error log is written, thus preventing the unit from being used under computer control until checked by an authorized person.

5.5. Communication Link Errors

Each record that is transmitted across the link between the two computers includes a checksum word; furthermore, all records are echoed back to the sender, which verifies that the echo matches the original. The link between the control computer and the treatment unit is a 16-bit duplex, i.e., 36 wires: regrettably, there is no parity bit. Noise is detected and eliminated as described earlier. An actual fault with the connection to the unit is detected by the form of the twitcher words. If there is an open circuit or a short circuit, then at least one of these words will be corrupted. Since the treatment unit does not receive the valid twitch, it stops the drives and closes the shutter.

5.6. Operator Errors

The computer will cause the treatment unit to perform exactly the same actions each time the patient is treated; there is no possibility of a collimator being set incorrectly or a wedge forgotten. However, if the treatment file is incorrect, then the incorrect treatment will be given every time. Although the editor program checks the reasonableness of a treatment file, it is impossible for it to fully verify that the treatment file is giving the dose intended by the doctor. At the Royal Free Hospital, treatment files are created by one physicist and then listed and independently checked by another physicist.

The radiographer must set up the patient correctly as in a normal treatment procedure; however, only the patient has to be set up; the computer sets the collimators and other machine parameters, so that a large group of errors, i.e., machine-settings errors, are eliminated. Since the computer can move the patient before opening the shutter for the first field, the initial set-up position does not require the target to be at the machine isocenter. It is often more convenient to set up at some other easily definable point, then ask the computer to move the target to the isocenter, thereby greatly reducing errors due to fatigue or forgetfulness.

Unfortunately, there is a design fault in the RFH treatment unit that can lead to errors. If the radiographer forgets to switch the unit over to computer control before pressing *treatment start*, then the unit acts as a normal manual unit; i.e., it opens the shutter, which may mean that the isocenter is not at the target. If this does occur, the computer detects the opening of the shutter, but since it does not have control, it cannot close the shutter. All the computer can do is repeatedly flash a message on the screen, causing the keyboard to bleep continually until the shutter is closed by pressing *treatment stop*. A radiographer's error, like all other errors, is logged and the exact exposure time is recorded.

6. FUTURE DEVELOPMENTS

Four developments are combining that will affect the future of computer-control treatment units. Firstly, the CT scanner is becoming a very common diagnostic aid, which is forcing radiotherapists to think of targets in three dimensions or at least as a series of two-dimensional images. The need to shape the high-dose volume in three dimensions, therefore, becomes self-evident. Secondly, the ubiquitous microchip is playing a more and more active role within conventional treatment units, and it should be a relatively easy task to interface with a treatment unit control circuit that uses computer chips. Thirdly, manufacturers are moving toward "assisted set-up" where, e.g., collimators are set automatically at the start of each field. The fourth factor is simply that the cost of computers is dropping rapidly, whereas their power and reliability is increasing rapidly. It seems inevitable, therefore, that computer-controlled teletherapy units will be more common in the future. It will be a pity if such units were restricted only to being able to move while there was no exposure; i.e., for assistive set-up and movement between fields, since this would not permit conformation therapy to be given.

All the forms of conformation therapy put a considerable demand on the planning physicist and dose calculation programs. At the Royal Free Hospital, we are studying how computers can assist in generating conformation treatment plans. The goal of a completely automatic planning system, where the target is outlined by the therapist and the computer produces the treatment file automatically, is probably not attainable for some time, since it has not yet been fully achieved for simple single-plane treatments. We have written an experimental program that does assist the physicist in producing a conformation treatment. Currently, patient outlines, inhomogeneities, and the target are digitized and fed into the computer. The data will eventually come directly from machine-readable CT scanner images on which the radiotherapist has "drawn" the target. The physicist must enter the prescription data, e.g., dose per treatment, the type of treatment (multiple track, etc.), and the longitudinal weightings for the received dose along the track and also the percentage dose to be given by track. From these data, the program calculates collimator sizes, couch position and speeds, and automatically generates a treatment file in about 10 min. The generated treatment files agree very well with those calculated manually, but more testing is needed before the system can be used for patients. Having created a treatment file, a dosimetry system is needed to calculate the dose given during the treatment. A typical multitrack treatment can be considered as a series of up to 300 separate fields, and the effect of each field on each slice must be calculated. At present, our dosimetry system uses a very simple off-axis algorithm, but it does enable us to determine that no gross errors have occurred. To calculate the dose distribution for a typical multitrack treatment over ten slices takes about 4 h of CPU bound computing.

REFERENCES

1. J. A. Brace, T. J. Davy, and D. B. L. Skeggs, A computer-controlled cobalt unit for radiotherapy, *Med. Biol. Eng. Computing* **19**, 612–616 (1981).
2. J. A. Brace, A computer system for the dynamic control of a tele-cobalt unit, in *International Symposium on Fundamentals in Technical Progress, Liege, Belgium, May 1979*, Vol. 2, Presses Universitaires de Liege, pp. 12.1–18 (1979).
3. J. A. Brace, A computer-controlled tele-cobalt unit, *Int. J. Radiat. Oncol. Biol. Phys.* **8**, 2011–2013 (1982).
4. A Chung-Bin and T. Wachter, Applications of computers in dynamic control of linear accelerators, in *Proceedings of the Third International Computer Software and Applications Conference 1979*, IEEE (1979), pp. 270–275.
5. B. Bjarngard, P. Kijewski, and C. Pashby, Description of a computer-controlled machine, *Int. J. Radiat. Oncol. Biol. Phys., Suppl. 2* **2**, 142 (1977).
6. M. B. Levene, P. K. Kijewski, L. M. Chin, B. E. Bjarngard, and S. Hellman, Computer-controlled radiation therapy, *Radiology* **129**, 769–775 (1978).
7. L. M. Chin, P. K. Kijewski, G. K. Svensson, J. T. Chaffey, M. B. Levene, and B. E. Bjarngard, A computer-controlled radiation therapy machine for pelvic and para-aortic nodal areas, *Int. J. Radiat. Oncol. Biol. Phys.* **7**, 61–70 (1981).
8. H. Perry, J. Mantel, and J. J. Weinkam, Optimized treatment planning: Automatic control of some parameters, in *International Symposium on Fundamentals in Technical Progress, Liege, Belgium, May 1979*, Vol. 2, Presses Universitaires de Liege, pp. 13.1–13.35 (1979).

2

Measurement of Human Body Composition *in Vivo*

LEWIS BURKINSHAW

1. INTRODUCTION

Until almost the middle of this century, the study of human body composition was the province of the anatomist and pathologist, whose methods of investigation were dissection and analysis of tissues postmortem. Two developments during the 1940s first made practicable the quantitative study of human body composition *in vivo*. The first was the enunciation by Behnke and his colleagues of the idea that the body is made up of two components, fat and lean tissue, whose proportions in an individual can be deduced from the measured density of the body.[1] The second was the increasing availability of radioactive isotopes, which can serve as tracers to determine the masses of body compartments by dilution. Behnke's method required the subject to be weighed under water, and it was therefore applicable only to people willing and able to be immersed; the dilution method made no such demands on the subject and opened up the study of the body's composition in patients with a variety of diseases.

During the last 40 years, ionizing radiations have been used to study the composition of the body in more and more detail. For instance, the mineral content of part of a bone can be measured by observing the attenuation of a beam of gamma rays passing through it;[2] the amount of potassium in the body can be estimated by counting gamma rays emitted by the natural

LEWIS BURKINSHAW • Department of Medical Physics, University of Leeds, The General Infirmary, Leeds LS1 3EX England.

radioisotope ^{40}K;[3] the amounts of five other elements (N, Na, Cl, P, Ca) can be determined by analyzing the gamma ray spectrum recorded after the body has been irradiated with neutrons.[4]

As techniques proliferate, it is important from time to time to consider as objectively as possible how far they advance our understanding of human biology in health and disease and to what extent each new technique is an improvement over existing methods. These are the aims of this chapter.

2. AIMS OF STUDIES OF HUMAN BODY COMPOSITION

Studies of human body composition are, in general, designed either to measure the composition of normal, healthy subjects or to assess departures from normality. Measurements of normal body composition have both fundamental and practical value. At the fundamental level, these measurements yield data about normal growth, maturity, and aging essential for developing theories about growth. At the practical level, they provide standards of reference against which to judge departures from normality. For both purposes, it is necessary to quantify differences between genetic groups, differences between sexes within each group, systematic variations with age and body size, and the distribution of those seemingly random differences between individuals that remain unexplained.

Studies of abnormal body composition may still involve measuring healthy people, for not all abnormalities, in the sense of significant deviations from the normal mean, are disadvantageous. Examples are studies of the body composition of athletes[5, 6] or people living in extreme environments.[7] Nevertheless, abnormalities of body composition are often associated with disease and may be studied either in the hope of elucidating disease processes or as a means of monitoring the progress of a disease or its response to treatment.

3. SELECTION OF ASPECTS OF BODY COMPOSITION TO BE MEASURED

As pointed out by Siri[8] in 1956, "the gross composition of the body can and has been described in a variety of ways, ranging from a tabulation of its elemental constituents to a description of the proportions of its specific organs." Since 1956, tabulations have become if anything more exhaustive, as instanced by the compilation of "Reference Man,"[9] and the detailed study of the elementary composition of individuals has been facilitated by the development of *in vivo* neutron activation analysis.[10] On the other hand, there have been no universally accepted advances in methodology for measuring the masses of organs and tissues *in vivo*.

The amount of detail required depends on the object of the investigation. If it is to accumulate data on normal subjects for heuristic purposes, then all information, in any degree of detail, is admissible. If the object is to study a specific abnormality, then an analysis should be chosen to suit the problem.

Elementary analysis of the body is seldom of interest in itself. The most important exception is perhaps body potassium, which has been extensively studied in investigations designed to reveal intracellular deficiency of the ion.[11] More often, body elements are measured as a means of estimating the mass of some functional component of the body. For instance, potassium serves this purpose also as a measure of the mass of fat-free tissue (Section 5.4).

For many purposes, it is highly informative to divide the body into fat and the fat-free tissues; that is, to estimate the sizes of the body's main energy store and its active, metabolizing component. This was first made practical 40 years ago when Behnke and his colleagues[1] developed whole-body densitometry into a workable technique, and a great deal of work on body composition has since been based on this two-component model.

Nevertheless, it is clearly a gross oversimplification to treat fat-free tissues as a single homogeneous entity. Even in healthy people, relative proportions of such tissues as skeletal muscle and bone are of interest, and any study of body composition in disease should be concerned with changes in the make-up, as well as the amount, of lean tissue.

In his 1956 review, Siri[8] pointed out that the chief functional constituents of the body are water, fat, protein, and minerals, which, apart from a few hundred grams of carbohydrate, and even smaller amounts of other organic substances, form the entire bulk of the body. He therefore considered that analysis of the body into these four components would be of immediate physiological and clinical interest. At the time, it was impossible to measure the masses of protein and minerals *in vivo*. However, as will be explained in Section 6, this obstacle has since been removed by the development of *in vivo* neutron activation analysis, and the idea has been revived and applied successfully to the study of nutritional problems in surgical patients.[12]

Fruitful though this approach may be, the information that it gives is limited, for water, fat, protein, and minerals function only when organized into tissues. An analysis that gave the masses and compositions of at least the major tissues that go to make up the total fat-free mass should be more informative. Some progress has been made: Total body calcium, and hence skeletal mass, can be measured by *in vivo* neutron activation analysis (Section 4.3), and a few methods of measuring the mass of skeletal muscle have been proposed (Section 7.2), though none has as yet been satisfactorily verified.

Thus, it is now technically feasible to analyze the living human body in a variety of ways. The present methodology of each approach, and possible future developments, will be reviewed in the following sections.

4. DETERMINATION OF THE ELEMENTARY COMPOSITION OF THE BODY

4.1. Analysis of Tissues at Biopsy or Postmortem

At first sight, it may seem inappropriate to discuss direct analysis of tissue samples, particularly of samples taken postmortem, in a review of methods of measuring human body composition *in vivo*. Nevertheless, although the measurements are not carried out on the intact human body, data acquired in these ways are invaluable aids in assessing and interpreting the results of less invasive methods.

In a few instances, whole body analyses by chemical means have been used to check results from other methods. To calibrate their method of measuring total body calcium by *in vivo* neutron activation analysis, Nelp *et al.*[13] irradiated and measured five cadavers, then related the observed counting rates to the masses of calcium in the cadavers, determined by ashing and chemical analysis. Chemical analyses of monkeys have been used to check a method of estimating total body potassium from measurements of natural radioactivity,[14] and analyses of pigs have been used to check the determination of body nitrogen from the measured intensity of prompt gamma radiation emitted during neutron irradiation.[15]

In a more general way, published analyses of tissues serve to check that new methods at least give results of the correct order of magnitude. Perhaps more importantly, these analyses provide the basis for interpreting *in vivo* measurements. For instance, tissue analyses have shown that water constitutes approximately 73% of the fat-free tissues;[16] since all the water in the body is in fat-free tissues, this factor is frequently used to estimate the fat-free mass of an individual from total body water measured by isotopic dilution (Section 4.2). Similarly, four published results of cadaver analyses gave the average potassium content of the fat-free body as 68.1 mmol.[17] This value is widely used to estimate fat-free mass from measured total body potassium (Section 5.4). Average values of the concentration of potassium in biopsies of skeletal muscle have been incorporated into models used to estimate muscle mass *in vivo* (Section 7.2).

Tissue masses estimated on the basis of such conversion factors can at best be only approximately correct, because the factors themselves are approximate average values. Widdowson[18] has outlined the difficulties of cadaver analysis, and many of the cadavers analyzed could scarcely be regarded as of normal composition. The composition of individual tissues is not universally constant; for instance, muscle biopsies give evidence of variations of electrolyte concentrations in wet muscle not only between individual subjects, but also between muscles in the same individual and between sites in the same muscle.[19, 20]

4.2. Isotopic Dilution

To measure the exchangeable mass of a body element by dilution, the patient is given a small dose T of a tracer, often a radioisotope of the element, either orally or intravenously. The tracer is allowed to reach a steady distribution within the body, and during the equilibration period the amount of tracer excreted E is measured. A sample of plasma or urine is then taken, and concentrations of the stable element m and tracer t in the sample are measured. Then the exchangeable mass of the element M is given by

$$M = \frac{T - E}{t} m \qquad (1)$$

For the dilution method to be successful, the administered tracer must mix rapidly and uniformly with the majority of the element to be measured, and only a small amount of tracer must be excreted during the mixing period. In practice, this limits the method to measuring the electrolytes potassium, sodium, and chlorine, and even the last of them is usually determined indirectly. For lack of a radioisotope of chlorine of convenient half-life, exchangeable chlorine is estimated by dilution of ^{82}Br (half-life 36 hr). Methods have been developed to the point where all three elements can be determined simultaneously with precision of the order of 3 to 4%,[21] and all three have been used extensively in clinical research.[22]

A practical disadvantage of dilution methods is that 24 hours must pass for the tracer to equilibrate, or up to 48 hours in disease.[21, 23] A more fundamental problem is that, even after apparent equilibrium has been reached, the tracer has not exchanged completely with all of the element, and, therefore, the measured exchangeable mass is less than the total body content.

In the case of potassium, Moore[24] is convinced that the discrepancy is less than 0.5%. However, direct comparisons between exchangeable and total body potassium, the latter measured by whole-body radiation counting (Section 4.3), do not support this conclusion. One comparison carried out in our laboratory,[25] and others cited by Belcher and Vetter,[21] suggest that the mean discrepancy is of the order of 5%, while others[23, 26, 27] indicate that, in some diseases, exchangeable potassium underestimates total body potassium by 30 to 40% at 24 hr and by 10 to 15% at 48 hr. Reviews of comparisons of exchangeable sodium with total body sodium determined by *in vivo* neutron activation analysis[28, 29] suggest that 20 to 30% of total body sodium is nonexchangeable.

4.3. Measurement of Natural or Induced Radioactivity

Potassium is unique among the essential body elements in that it emits penetrating gamma rays that can be measured by radiation detectors placed around the body. Gamma rays, of energy 1.46 MeV, are emitted following the decay of the natural radioisotope of potassium ^{40}K, which is present in constant proportion in all natural potassium. Therefore, the total mass of potassium in the body can be estimated from the measured intensity of the emitted gamma radiation. This technique originated in our laboratory[30] and as sensitive whole-body radiation counters have become available, has been developed here and elsewhere into a clinically acceptable procedure capable of measuring total body potassium with a standard error of 3 or 4%.[3]

The principle of this technique has been applied to measuring other elements by activating them artificially. These developments and their present status are fully described elsewhere;[10] therefore, only the principles of the most widely used method, *in vivo* neutron activation analysis, are given here.

When the body is irradiated with fast neutrons, penetrating gamma rays are emitted during irradiation and for some time afterward. These gamma rays are products of interactions between neutrons and atomic nuclei in the body, and their energies are characteristic of the nuclei that emit them. Therefore, if energy-sensitive radiation detectors are placed around the body, a complex spectrum of gamma ray energies is acquired that can be analyzed to give the total amounts of some elements in the body. In order of abundance, the essential elements that can at present be measured are oxygen, carbon, hydrogen, nitrogen, calcium, phosphorus, sodium and chlorine.

We at Leeds have devised a clinically acceptable system that simultaneously estimates the total body contents of potassium, nitrogen, calcium, phosphorus, sodium and chlorine, with accuracies of a few percent, by analyzing the gamma ray spectrum acquired after irradiation with 14-MeV neutrons.[4] We also originated a method of estimating total body carbon *in vivo* by counting the 4.43-MeV gamma rays emitted during irradiation as a result of inelastic scattering of fast neutrons by carbon nuclei.[31] With our present prototype apparatus, carbon has to be determined separately from the other six elements. However, we are currently developing new apparatus that will estimate all seven elements with one irradiation of the patient. The analysis will then be extended to estimate hydrogen from the measured intensity of its neutron capture gamma rays and oxygen from the activity of ^{16}N created by the reaction $^{16}O(n, p)^{16}N$.[32]

The conspicuous advantage of nuclear-activation techniques over isotopic dilution (Section 4.2) is that, unless specifically designed to measure only part of the body, these techniques unequivocally measure the total-body content of each element. Their most frequent clinical applications to date have been in measuring nitrogen, and, hence, protein (Section 6) and calcium, as a measure of bone mineral mass (Section 7.1).

5. ESTIMATION OF THE MASSES OF FAT AND FAT-FREE TISSUE

5.1. Estimation from Body Density

This method, first established as a reliable technique for human studies 40 years ago by Behnke, Feen, and Welham[1] is still widely regarded as the most absolute way of measuring body fat. The method relies on the assumption that fat and the fat-free tissue have different but constant densities, in which case it can be shown that the fraction of body weight that is fat (f) is related to body density (ρ) by the equation

$$f = \frac{4950}{\rho} - 4.50 \tag{2}$$

The numerical values in this equation are correct if the densities of fat and fat-free tissue are 900 and 1100 kg/m^3, respectively,[33] and ρ is expressed in kg/m^3. Other values have been proposed,[8, 34] but they seem to have no better claim to accuracy than the round figures quoted. The value 900 kg/m^3 is the measured density of human fat at 37°C;[35] the value 1100 kg/m^3 was calculated by Siri[33] for lean tissue composed of 72% water, 21% protein, and 7% mineral, and it is consistent with the report by Behnke, Osserman, and Welham[36] that they had never found measured body density to exceed this value even in their leanest subjects.

Body density is found by dividing measured body weight by body volume. Body volume is most commonly determined according to Archimedes' principle by weighing the subject in air and under water and correcting for the volume of air in the lungs. This technique has been reviewed by Siri.[8] It is capable of high precision: published standard deviations of repeated measurements of body density in subjects of presumed constant composition range from 0.8 kg/m^3[37] to 4.3 kg/m^3,[38] all less than 1% of the measured density. A typical value is 2.3 kg/m^3.[39] Suppose that repeated measurements of the density of a subject show this standard deviation and a mean of 1055 kg/m^3, corresponding to a fat content of 19.3%

(13.5 kg if body weight is 70 kg, i.e., the composition of Reference Man[9]). Then it can be shown that the standard deviation of the fat values will be 1.02% of body weight, or 0.72 kg of fat, i.e., 5.3% of the subject's fat content. Thus, a precision of 0.2% in determining body density results in a precision of the order of 5% for body fat; the fat-free mass, as the larger component, is determined with better precision (1.3% in this example). In these calculations, as in other calculations of errors in this section, errors in measuring body weight are neglected. Even including true variations in body weight, the coefficient of variation of repeated measurements should not exceed 0.5%.[40]

The densitometric method has been widely used to analyze the composition of healthy subjects, but it has no place in clinical investigations. It is clearly impracticable to immerse ill patients in water, and alternatives to underwater weighing, such as measuring the displacement of helium gas,[8] have not been developed into workable clinical procedures. Furthermore, the accuracy of the method will be impaired if the composition (and, hence, the density) of the fat-free tissues is changed by disease.

Such a change is likely to occur, for instance, when patients lose weight following a surgical operation. Kinney et al.[41] found that, during the first ten days after major surgery, male patients lost on an average 6% of their body weight and that the tissues lost were composed of 13% fat, 77% water, and 10% protein. Assuming that the loss of minerals over this short period is negligible, the patient is left with fat-free tissues containing relatively less water but more protein and minerals than before; the average density of the fat-free tissues is therefore increased.

To estimate the magnitude of the effect, consider a hypothetical male patient whose initial body composition is as shown in the first column of Table 1. This patient has the body weight and fat content of Reference

Table 1. Initial and Final Composition of a
Hypothetical Surgical Patient[a]

	Initial mass	Mass loss	Final mass
Water	40.9	3.2	37.7
Protein	12.0	0.4	11.6
Minerals	3.6	0.0	3.6
Fat-free	56.5	3.6	52.9
Fat	13.5	0.6	12.9
Total	70.0	4.2	65.8
Density	1055	—	1059

[a] The patient is assumed to lose tissue of the composition reported by Kinney et al., 1968. Body densities are calculated assuming the densities of fat, protein, and water given by Siri, 1956; the initial density of the fat-free mass is 1100 kg/m³. Masses are given in kg and densities in kg/m³.

Table 2. *True and Calculated Body Composition of the Hypothetical Patient in Table 1*[a]

	Final values				Mass losses			
	Mass		Error				Error	
	True	Calculated	(kg)	(%)	True	Calculated	(kg)	(%)
Fat-free	52.9	54.3	+1.4	+3	3.6	2.2	−1.4	−39
Fat	12.9	11.5	−1.4	−11	0.6	2.0	+1.4	+233

[a] The calculated values are derived from the final body density using Eq. (2). Masses are given in kg.

Man,[9] and the amounts of water, protein, and minerals are such as to give the fat-free tissues a density of 1100 kg/m³, assuming that water, protein, and minerals have densities of 993, 1340, and 3000 kg/m³.[33] Then, assuming that fat has a density of 900 kg/m³,[33] an accurate measurement of body density would give a value of 1055 kg/m³, which when substituted into Eq. (2) would estimate correctly the proportion of fat in the body.

Now, suppose that the patient loses 6% of body weight (4.2 kg) containing fat, water, and protein in the proportions found by Kinney *et al.*[41] The composition of the material lost and the final composition of the body are given in Table 1. With this final composition, an accurate measurement of body density would give a value of 1059 kg/m³. However, the density of the fat-free tissues is now 1106 kg/m³ rather than 1100 kg/m³ as was assumed when Eq. (2) was formulated. Therefore, when the final body density (1059 kg/m³) is substituted into Eq. (2), the proportion of fat in the body is estimated inaccurately. Table 2 shows that the error is equivalent to 1.4 kg of fat, i.e., 11% of the fat content and 3% of the fat-free mass. Table 2 also shows that the loss of fat is overestimated by more than 200% and the loss of fat-free mass is underestimated by 39%.

This example shows that the accuracy with which body fat and changes in body fat can be estimated from body density is impaired by typical pathological changes in the composition of the fat-free tissues. There would seem to be little point in trying to adapt the technique for clinical use.

5.2. Estimation by Anthropometry

Because body density can be measured successfully only by skilled operators working under laboratory conditions with motivated subjects, simpler alternatives have been devised. The aim is to estimate body density, and hence total body fat content, by measuring the thickness of subcutaneous adipose tissue, with or without measuring body dimensions. To establish such a technique, the chosen thicknesses and dimensions, and body density, are measured in groups of suitable subjects. The results are expressed as mathematical relationships between body density and the other measured quantities, which are then used to estimate body density for subjects whose density has not been measured directly.

Lohman[42] has recently reviewed techniques for estimating body fat from skinfold thicknesses. His survey of 23 samples of subjects gave a pooled standard error of estimated density of 7.0 kg/m^3 for young adult males, or 9.1 kg/m^3 for all other subjects. Applying the smaller value to the hypothetical subject of density 1055 kg/m^3 (Section 5.1), the standard error of the fat content estimated from Eq. (2) is 3.1% of the body weight of 70 kg, i.e., 2.2 kg of fat; this is 16.1% of the mass of fat in the body. Once again, the fat-free mass, as the larger component of body weight, is estimated with a smaller relative error; in this case, 3.9%. Thus, these methods give only rough estimates of the masses of fat in individuals but should accurately estimate mean values for groups of healthy subjects.

Systematic differences are found between skinfold measurements made by different operators on the same subjects. In our laboratory, using the method of Durnin and Womersley,[43] we found differences equivalent to a difference of 4% in the mass of fat in the body of a typical man,[44] i.e., 0.5 kg for Reference Man.[9] This is less than the standard error of the estimated fat content of an individual (2.2 kg).

The precision of this type of method is high. Hill et al.,[12] also using Durnin and Womersley's method,[43] estimated from the observed precision of the skinfold measurements that body fat was determined with a precision of 0.3 kg. In an unpublished study in this hospital, Blackett measured the skinfold thicknesses of ten patients three times in one day and found the precision of their estimated fat contents to be 0.22 kg, or 2.4% of the mean value. These figures compare favorably with the precision obtainable when body density is measured directly (0.7 kg; Section 5.1).

Figures in this section for the accuracy with which fat and fat-free mass can be estimated from measured skinfold thicknesses apply only if the subjects measured are typical of the groups of healthy subjects whose data were used to establish the method. The accuracy of estimation for subjects of abnormal body composition is uncertain; our own data suggest that it may be similar to that for normal subjects. In one study, fat-free mass estimated from skinfold thicknesses in 24 surgical patients and nine healthy volunteers was compared with total body nitrogen measured by neutron activation analysis in the same subjects.[45] The two variables were highly correlated, and their values were consistent with a mean concentration of nitrogen in the fat-free tissues of 34 g/kg, the value found by Widdowson[18] from cadaver analysis.

In a further study,[46] we analyzed data from 81 surgical patients (40 women and 41 men) who had lost, on average, 6 kg (10%) of body weight. Fat-free mass was estimated in three ways: (1) from skinfold thicknesses,[43]

(2) from total body nitrogen (assuming 34 g of nitrogen per kg), and (3) from total body water (assuming 730 ml water per kg; Section 5.3). The data were analyzed by factor analysis to estimate (1) the linear relationships between the three measures and (2) the component of the variance of each that was not associated with variations in the other two; the latter was taken to be an estimate of the random error variance of each method (Section 8.2). The slopes of the relationships were found to be close to unity, and the standard error of fat-free mass derived from skinfold thicknesses was found to be 3.2 kg, in reasonable agreement with the standard error of estimate of 2.2 kg given earlier in this section. Thus, in these patients, fat-free mass estimated from skinfold thicknesses was related to other measures of the same quantity, and had approximately the same standard error, as would be expected in healthy people. This finding suggests that the accuracy of the skinfold method was not seriously impaired by disease and loss of weight.

In summary, skinfold-thickness measurements are a relatively simple means of estimating fat and fat-free mass; the estimates they give are precise but not very accurate for an individual subject. There is some evidence that the methods are not markedly less accurate for wasted patients, but more thorough studies are required, including investigations of the effects of specific abnormalities, such as edema.

5.3. Estimation from Total Body Water

Fat is anhydrous, and, therefore, all the water in the body is contained in the fat-free tissues. If the proportion of water in the fat-free tissues is assumed constant and is known, then fat-free mass can be calculated directly from measured body water.

Total-body water can be determined by dilution (Section 4.2). According to a review by Sheng and Huggins,[47] the first measurements were made 50 years ago, using as tracer water in which one of the hydrogen atoms was replaced by deuterium. During the late 1940s and early 50s, such compounds as antipyrene enjoyed a brief vogue as tracers, but at present, the most commonly used tracer is water labeled with tritium, a β-emitting radionuclide that can be assayed accurately by liquid scintillation counting. The analytical standard error of this technique is approximately 1 l.[21, 48]

There is some evidence that dilution of labeled water overestimates body water because of the exchange of the label with nonaqueous hydrogen; in some species of animals, the discrepancy may be as much as 15% of body weight.[47] The size of the error in humans is unknown, but it has been estimated to lie between 0.5 and 2.0% and is usually neglected.

Assuming that body water can be measured accurately, a value for the proportion of water in the fat-free tissues is required. A value of 73.2%, based on analyses of animals by Pace and Rathbun,[16] is often used, but in a survey of eight published analyses of adult human cadavers, Sheng and Huggins[47] found values ranging from 70.1 to 82%, with a mean of 74.8%. If the fat content of each cadaver was estimated using the value of 73.2%, mean estimated fat content was 1.9 kg (21%) less than mean measured fat content.

Thus, the accuracy with which fat and fat-free mass can be estimated from measured body water is uncertain. The analytical standard error (1 l of water) is equivalent to 1.4 kg of fat-free mass or fat, i.e., 2.5% of the fat-free mass and 10% of the fat content of Reference Man.[9] Factor analysis of data from 81 surgical patients[46] (Section 5.2) suggests that values for fat-free mass derived from body water contain more random error than values determined from skinfold thicknesses or total body nitrogen. The standard errors were 5.5 kg, 3.2 kg, and 3.4 kg, respectively.

5.4. Estimation from Total Body Potassium

Potassium is another constituent of fat-free tissues that is not present in fat. Therefore, as with water, if its average concentration in the fat-free tissues is a known constant, fat-free mass can be determined directly from total body potassium. As explained in Section 4.3, total body potassium can be determined with a standard error of about 4% by measuring the natural radioactivity of the body.

The choice of a value for the average concentration of potassium in the fat-free tissues has been the subject of controversy[49, 50] that is still unresolved. The value of 68.1 mmol/kg, derived by Forbes, Gallup, and Hursh[17] from analyses of one female and two male cadavers, together with an earlier result from the literature, has been adopted by many workers. However, a review of published measurements of body potassium and fat-free mass in vivo[50] gives average values about 3% lower for men and 15% lower for women. Forbes himself has proposed a value 5.8% lower (64.15 mmol/kg) for adult women,[51] and there is evidence that the concentration is still lower in older people.[52] A study of our own and other workers' data, suggests that, regardless of age or sex, the concentration tends to be higher in people with higher fat-free mass.[81]

The analytical error in measured body potassium is equivalent to a standard error of 2.2 kg of fat-free mass or fat (assuming 3580 mmol of potassium in 56.5 kg of fat-free tissue[9]); this is equivalent to 4% of the fat-free mass and 16% of the fat content of Reference Man.[9] However, because of the uncertainties just mentioned, masses of fat and fat-free tissue can be determined only approximately from total body potassium even in healthy subjects, and values found for patients, whose potassium status may be abnormal, are presumably even less reliable. It might be better to interpret total body potassium as a measure of something other than the total fat-free mass: Moore and his colleagues[22, 24] use total body potassium as a measure of cell mass, and since skeletal muscle contains a higher concentration of potassium than most other tissues,[9] it figures in several proposed methods for estimating muscle mass (Section 7.2).

5.5. Estimation of Total Body Fat by Dilution

In all the methods described in Sections 5.1–5.4, the mass of one component is found by subtracting the mass of the other from body weight. Therefore, assuming that body weight is measured with negligible error, the errors in the two components, in units of mass, are equal and opposite, and the percentage error in body fat, the smaller component, is larger than that in fat-free mass. In consequence, when fat-free mass is estimated from total body water or potassium (Sections 5.3 and 5.4), the percentage error in the measured fat content is predictably greater than the percentage error in the actual measurement by a factor of four for a subject with the composition of Reference Man.[9] A better alternative would be to measure the body content of some substance present in only fat; then body fat would be estimated with the percentage error of the measurement technique, and fat-free mass with a smaller percentage error.

Fat has no unique natural constituent that can be measured *in vivo*, but certain substances are much more soluble in fat than in other body tissues and can therefore, in principle, be used to measure total body fat by dilution. Techniques have been developed using fat soluble gases, usually cyclopropane or radioactive ^{85}Kr,[53, 54] which are capable of precisions of 6 to 8% (about 1 kg of fat for Reference Man[9]). These methods seem unlikely to be used in general, since they require the subject to wear a mask or helmet continuously for periods from 2 to 9 hr.

5.6. Estimation of Body Fat from Total Body Carbon, Nitrogen, and Calcium

Although fat has no unique natural constituent that is measurable *in vivo*, it does contain a higher concentration of carbon than do fat-free tissues. Therefore, we have developed a technique for estimating total body fat from measured total body carbon, nitrogen, and calcium.

Total body carbon is determined by counting the 4.43-MeV gamma rays emitted when fast neutrons are scattered inelastically by carbon nuclei.[31] Total body nitrogen and calcium are determined by *in vivo* neutron activation analysis.[4] Body carbon is assumed to be contained entirely in fat, protein, and bone mineral, forming 77% of fat and 52% of protein and being present in bone mineral in the proportion of 740 g/kg of calcium.[9] Body protein is calculated by multiplying total body nitrogen by 6.25, assuming that 16% of the mass of protein is nitrogen and the body contains a negligible amount of nonprotein nitrogen. All the calcium in the body is assumed to be in bone mineral. These considerations lead to the formula

$$TBC = 0.77\,TBF + 3.25\,TBN + 0.74\,TBCa \qquad (3)$$

where TBC, TBF, TBN, and TBCa denote total body carbon, fat, nitrogen, and calcium, respectively. Hence,

$$TBF = 1.30\,TBC - 4.22\,TBN - 0.96\,TBCa \qquad (4)$$

From the observed precisions of the measured quantities (i.e., the standard deviations of repeated measurements of an anthropomorphic phantom), the precision of the body fat of Reference Man,[9] calculated from Eq. (4) is 0.66 kg, or 4.9%. The precision of fat-free mass is 1.2%.

At this early stage of development, the accuracy of this method is uncertain. If it is equal to the precision, then the method is as accurate as densitometry (Section 5.1) and considerably more accurate than the other three methods described in Sections 5.2–5.4. Because this method relies on measuring the basic chemical constituents of the body, it is likely to be more accurate than alternative methods for patients with abnormal tissue composition.

6. ANALYSIS OF THE FAT-FREE MASS INTO WATER, PROTEIN, AND MINERALS

The methods discussed in Sections 5.1–5.4 are all to some extent vitiated by uncertainties in the composition of fat-free tissues. Therefore, there is a strong case for measuring the primary chemical components of the fat-free tissues: water, protein, and minerals. This breakdown is potentially more informative than a simple estimate of total fat-free mass, and the latter can be found, if required, by adding the masses of the components.

A composite technique has been described by Hill et al.,[12] who determined protein and minerals by in vivo neutron activation analysis. Protein was calculated by multiplying total body nitrogen by 6.25 (Section 5.6), and the mineral content was approximated by the sum of the masses of potassium, sodium, chlorine, phosphorus, and calcium. Water was not measured directly; instead, fat was deduced from measured skinfold thicknesses,[43] and water content was found by subtracting protein, minerals, and fat from body weight. In a subgroup of patients, the researchers also measured body water by dilution and found satisfactory agreement between body weight and the sum of the four components.

We have found this to be a fruitful way of studying the body composition of surgical patients, but only because we have in vivo neutron activation analysis available to measure body protein directly. In principle, this important component could be found as a difference if the other three were measured, but since protein forms only about 15% of the total body,[9] any such estimate would be very inaccurate.

Given that protein and minerals can be measured, which of the other two components should be measured? An obvious answer is to measure both and compare the sum of the four components with body weight as a check on the overall accuracy of the techniques. Otherwise, the author's preference is to measure body water, which seems more likely to be correct when body composition is abnormal than commonly available assays of total body fat (Sections 5.1–5.4). Furthermore, body water can often be measured when body fat cannot; thus, Siwek,[55] by subtracting water, protein, and minerals from body weight, was able to determine body fat in patients so obese that neither body density nor skinfold thicknesses could be measured. If this method is applied to Reference Man,[9] body fat and fat-free mass (the sum of water protein and minerals) are estimated with a standard error of 1.3 kg, which is 10% of body fat and 2.3% of the fat-free mass.

7. ESTIMATION OF THE MASSES OF INDIVIDUAL TISSUES

7.1. The Skeleton

The body contains approximately 5 kg of bone mineral, forming about 9% of the total mass of fat-free tissue.[9] The mineral contains practically all of the body's calcium; therefore, the most direct way of estimating total bone mineral is to measure total body calcium by *in vivo* neutron activation analysis.[10] The measurement has a precision of 2 to 3% and an accuracy of about 4%.[56]

For many purposes, particularly in sequential studies of pathological changes in bone mineral, nothing is gained by converting measured total body calcium into an estimate of total bone mineral. However, it is useful to make this conversion when trying to estimate proportions of the tissues that make up the fat-free mass. In such a context, Cohn et al.[57] divided total body calcium by 0.208 to give bone mineral mass, with the factor derived from the work of Bigler and Woodward.[58]

When facilities for measuring total body calcium are not available, its value for a normal subject can be estimated from more easily measurable quantities, such as height, span, years after menopause (for women),[59] and total body potassium.[60] The standard error of the estimate is typically 7 or 8%. Correlation coefficients of 0.8 to 0.9 have been found between total body calcium and bone mineral content measured at specific sites on long bones by gamma ray absorptiometry,[61] suggesting that such local measurements could be used to estimate total body calcium. We investigated this possibility by measuring a group of 42 patients with diseases affecting bone mineral content. We found that total body calcium content could best be estimated from a linear combination of the bone mineral contents of the femur and radial shaft. The standard error of the estimate was 11% of the mean calcium content. We therefore concluded that the only way of determining accurately the total calcium content of an individual patient was to measure it directly.[62]

7.2. Skeletal Muscle

Skeletal muscle makes up approximately half of the total mass of fat-free tissue in the adult human body.[9] Its most obvious function is giving the body movement, but it also serves as a reservoir of protein, which is used as a source of energy when fat stores run low. Therefore, it would be useful for both physiologists and clinicians to be able to measure the mass of skeletal muscle *in vivo*.

Several methods have been proposed. Perhaps the simplest method for clinical use, but the least direct, is estimating muscle mass from measured daily output of creatinine in the urine. Creatinine is a product resulting from the breakdown of creatine. Ninety-eight percent of the body's creatine is located in skeletal muscle, and approximately 2% of it is converted into creatinine per day.[63, 64] Therefore, assuming that all the creatinine is excreted in the urine, its rate of excretion is proportional to muscle mass. The exact value of the constant of proportionality is uncertain. Graystone[63] adopts a value of 20 kg of muscle per gram of creatinine excreted per day. Others[22] prefer a formula relating muscle mass to the creatinine coefficient, defined as the mass of creatinine excreted per kg of body weight per day. The accuracy of the method is not well established. Graystone[63] emphasizes the need to control dietary intake of creatine and protein, and Moore *et al.*[22] state that the excretion rate of creatinine is altered by recent severe injury or renal disease, casting doubt on the value of the method for clinical studies.

If it were possible to introduce a tracer into the body that exchanged uniformly with all the creatine in the muscle, and if the concentration of the tracer in the muscle could be measured, then the muscle mass could be estimated by dilution (Section 4.2). Two groups of workers have tried this approach. Kreisberg, Bowdoin, and Meador[65] injected creatine labeled with ^{14}C into two male and two female hospital patients and then measured the concentration of ^{14}C in a biopsy of the quadriceps muscle eight days later. They found muscle masses ranging from 50 to 67% of the fat-free mass. Picou *et al.*[64] used ^{15}N-labeled creatine in a similar way to measure muscle masses of eight male infants who had recovered from protein-energy malnutrition. They found muscle masses ranging from 1.0 to 4.9 kg.

It is difficult to assess the accuracy of this method, except to say that the few results reported seem plausible. The assumption that the muscles are uniformly labeled has not been verified in man, and Greatrex *et al.*[66] claim that it is not valid for dogs. They measured specific activities in muscles five to 17 days after administering labeled creatine and found values up to three times higher in some muscles than in others. This method does have the distinction of being more applicable to patients than to normal subjects, since few healthy people would volunteer for muscle biopsy, but it might often be possible to use part of a biopsy taken in the course of a patient's management.

A distinguishing feature of skeletal muscle is that it contains a higher concentration of potassium than do most other tissues. Therefore, the average concentration of potassium in the fat-free tissues should be higher in muscular individuals than in less muscular ones.[67] This idea is the basis of two proposed methods for estimating muscle mass. The first method was proposed in 1963 by Anderson.[68] He assumed that the body is made up of skeletal muscle, containing potassium and water in the ratio 118 mmol/kg, and "muscle-free lean" and adipose tissue, each containing 75 mmol of potassium per kg of water. With these assumptions, he derived formulas giving the mass of each constituent as a function of measured values of body weight, total body potassium, and total body water. He estimated that muscle mass could be determined with a standard error of 11% (3.1 kg for Reference Man[9]).

More recently,[69] we suggested an alternative method based on the assumption that the body is made up of skeletal muscle containing potassium and nitrogen in the ratio 3.03 mmol/g, with "nonmuscle" containing 1.33 mmol of potassium per gram of nitrogen and fat containing neither potassium nor nitrogen. With these assumptions, we found that the mass of muscle in the body (M_m, kg) could be calculated by substituting measured total body potassium (TBK, mmol) and total body nitrogen (TBN, g) into the formula

$$M_m = (TBK - 1.33\ TBN)/51.0 \tag{5}$$

From our knowledge of the errors of measurement, we estimated that muscle mass was given with a standard error of 4 kg (14% for Reference Man[9]).

In Anderson's method and ours, body water and body nitrogen effectively serve to estimate fat-free mass. Therefore, any other estimate of fat-free mass can be used to give a formula similar to that in Eq. (5). Assuming that muscle contains 91 mmol of potassium per kg and nonmuscle 48 mmol/kg,[69] we find that

$$M_m = (TBK - 48\ FFM)/43 \tag{6}$$

where FFM is the fat-free mass in kg. The accuracy of this estimate of muscle mass depends on the method used to measure fat-free mass. If an anthropometric method is used, the standard error of the measured fat-free mass will be approximately 2.2 kg (Section 5.2). Assuming that total body potassium is measured with a standard error of 4%,[69] i.e., 143 mmol for a person with the composition of Reference Man,[9] the standard error of the estimate of muscle mass given by Eq. (6) is 3.5 kg (12%).

Anderson's method, our method, and their generalization all estimate muscle mass with similar errors. The errors are rather large because, as Anderson[68] points out, errors of measurement are magnified by the calculations: muscle mass is estimated by a relatively small difference between measured values (Eqs. 5 and 6). Furthermore, the errors given here will be enhanced by uncertainties in the assumed compositions of the tissues.[69] Therefore, when describing our method,[69] we suggested it would be useful for groups of subjects rather than individuals. However, Cohn et al.[57] claim that their method of measuring total body nitrogen is sufficiently accurate to remove this limitation.

The validity of this approach in estimating muscle mass awaits confirmation. Lukaski et al.[70] studied a group of 14 healthy young men and found that the rate of excretion of 3-methylhistidine, a product from the breakdown of muscle protein, was highly correlated ($r = 0.91$) with estimated muscle mass but not significantly correlated with nonmuscle mass. It may be that muscle mass tends to be underestimated: Cohn et al.[57] found that muscle formed only 35–40% of the fat-free mass of young and middle-aged healthy men, compared with a widely accepted value of 50%.[9] We have found occasional negative values for muscle mass in emaciated surgical patients.[69] Although the latter finding is not unexpected in view of the relatively large error in individual estimates, it is a salutary reminder that even if the method is valid for healthy subjects, it may be invalidated by disease, since cellular depletion of potassium will be interpreted as a loss of muscle.

8. FUTURE DEVELOPMENTS

8.1. Techniques of Measurement

The foregoing review suggests that, even after 40 years of development, our most urgent need is an improved method of measuring total body fat, valid for both healthy subjects and patients whose body composition has been altered by disease or treatment. This quantity is of prime importance in any study of nutrition and yet, even under laboratory conditions with healthy subjects, we seem unable to measure total body fat with precision better than about 0.7 kg, or 5% (Section 5). This precision is adequate for surveys but inadequate when, for example, we want to measure a change of 1 kg or so in the fat content of an individual, such as we have found in surgical patients fed intravenously for two-week periods.[71] Skinfold thicknesses may give better than 5% precision (Section 5.2), but they are less accurate. Furthermore, all the methods in Section 5, with the possible exception of the total body carbon method (Section 5.6), rely on assumptions about the composition of the lean tissue that may be invalid in disease.

Next in importance, in the author's opinion, is the need to improve our ability to measure the mass of skeletal muscle. This quantity is important in studies of growth, physical performance, and nutrition, yet none of the methods reviewed in Section 7.2 can be regarded as satisfactory.

The contribution of ionizing radiations to the study of body composition has reached something of a plateau now that *in vivo* neutron activation analysis has become an acceptable clinical procedure.[10] Nevertheless, some development is still possible: In our institution, we are currently developing a method for measuring total body carbon in the expectation that it will lead to accurate and precise measurements of body fat based on assumptions that are not invalidated by disease (Section 5.6); in Sweden, Ericsson[72] has measured total body hydrogen as a means of estimating "soft fat-free solids."

Future advances may well stem from other branches of physics. For example, the technique of nuclear magnetic resonance can be used to measure the concentration and spin-lattice relaxation time of water protons in tissue. These quantities vary between the tissues of the body, and apparatus has been developed that can show their distributions within cross sections of the body as tomographic images.[73, 74] The method has also been used to study the biochemistry of phosphate compounds in muscle.[75, 76] Understandably, imaging and studying the biochemistry of tissues *in vivo* have so far been regarded as the most profitable clinical applications of nuclear magnetic resonance, and it has yet to be applied to the quantitative study of gross body composition.

Other properties of tissues might be considered. For instance, tissues vary in their electrical conductivity and permittivity[77] and attempts have already been made to image distributions of these quantities.[78] Apart from any fundamental advantages they might have, methods that did not have the risk, however small, that accompanies using ionizing radiations would have the great practical advantage of being applicable to children and healthy adults as well as to hospital patients.

8.2. Evaluation of Techniques

It is extremely difficult to assess the accuracy, as opposed to the precision, of techniques for measuring body composition. The obvious solution of measuring cadavers and analyzing them chemically is so laborious that it is not widely practiced. Even if it were, such large-scale chemical analyses could not automatically be assumed to be more accurate than the method being assessed. The alternative of analyzing animals may be unsatisfactory, particularly with such methods as our implementation of *in vivo* neutron activation analysis,[4] which has been accurately calibrated only for the range of body sizes and shapes encountered in clinical research.

Usually, therefore, a new technique is assessed by using it to measure a group of subjects and comparing the results with those given for the same subjects by one or more established methods. If the results are well correlated, it can be concluded at least that the methods are responding together to differences between the subjects rather than varying randomly.

If three or more methods are used, the analysis can be taken further by means of factor analysis, assuming a single factor.[79] This procedure breaks down the variance of results from each technique into two components: (1) the part that is in common with the variability from the other sets of results and (2) the part that is independent. The former reflects true variations between subjects in the quantity being measured; the latter is an estimate of the random error of measurement. This analysis was proposed by Barnett[80] as a means of comparing laboratory instruments, although he did not refer to it as factor analysis in his paper.

In addition to analyzing the variance from each set of results, factor analysis also gives estimates, unbiased by errors of measurement, of the parameters in the assumed linear relationships between the measured quantities. These relationships cannot reveal systematic errors common to all the measured quantities, but they can be compared with the relationships to be expected on theoretical grounds.

Thus, factor analysis offers a way of extracting more information from the variances and correlation coefficients that are normally calculated from the results of an intercomparison. The main advantage of factor analysis is that it estimates the random error of each measurement technique from data acquired in the course of actually measuring real subjects rather than in specially contrived laboratory experiments.

In many investigations, it is necessary to measure changes in body constituents. It might be assumed that the most precise method would measure changes best, but this is not necessarily so. For example, the accuracy with which a change in body fat can be estimated by measuring skinfold thicknesses must depend on what proportion of the total change takes place in the subcutaneous fat layer. This aspect of evaluating techniques has been rather neglected in the past and should be given more attention in the future.

8.3. Interpretation of Data

As we have seen, few of the measurements we make when analyzing the composition of the living human body are direct measurements of body constituents. Usually, data are entered into some mathematical formula to give the mass of the constituent of interest, and the process of developing useful formulas has introduced simplifying concepts.

Of these, the most influential has undoubtedly been the concept of the "lean body mass" or "fat-free mass," which assumes that the fat-free tissues of the body (with or without a small quantity of essential lipid[36]) have the same average density in all subjects. This simplifying concept opened the way to modern studies of body composition, and it still may be useful to discuss normal body composition in these terms. However, now that the effects of disease on body composition are being investigated, by methods unrelated to densitometry, the concept of a fat-free mass invariant in composition is a constraint rather than a help. It would be preferable to measure more basic constituents, such as water, protein, and minerals (Section 6), which is no problem for investigators equipped to carry out multi-element neutron activation analysis *in vivo*. For those not so equipped, the close correlation that we found in surgical patients between total body nitrogen and fat-free mass derived from skinfold thicknesses (Section 5.2) suggests that the latter can serve as a measure of body protein even when body composition is affected by disease.

This criticism of the concept of the lean body mass applies equally to the more recent concept of "muscle-free lean"[68] or "nonmuscle" tissue,[69] and it is to be hoped that these simplifications will also be made unnecessary by more direct ways of measuring such important tissues as muscle. Nevertheless, although such concepts as these may be overtaken by advances in methodology, they will have to be replaced by other ways of interpreting acquired data, and it is imperative that conceptual development should keep pace with technical progress.

ACKNOWLEDGMENTS

This chapter could not have been written without the help and support of many people over a period of years. It is through the initiative of Professor F. W. Spiers that we are now able to measure body elements by *in vivo* neutron activation and whole-body radiation counting. Dr. C. B. Oxby has carried through the scientific development of this method with the able assistance of our technical staff and research students. Professor D. B. Morgan and Professor G. L. Hill not only involved us in stimulating clinical research, but spent many hours discussing the problems encountered in interpreting our data. Throughout this period, the Medical Research Council, my employer, has supported us with a series of grants.

Special mention must be made of the late Professor R. E. Ellis, a former member of the editorial board of this series, who, until his untimely death in

1981, was head of this department. When he joined us in 1972, he immediately saw the potential of our early attempts at neutron activation analysis *in vivo* and spared no effort to ensure that our research group remained intact through a period of insecurity. In the ensuing years, he continued to do everything he could to provide us with staff and material resources and to encourage our progress. We owe him a deep debt of gratitude.

REFERENCES

1. A. R. Behnke, B. G. Feen, and W. C. Welham, The specific gravity of healthy men. Body weight ÷ volume as an index of obesity, *J. Am. Med. Assoc.* **118**, 495–498 (1942).
2. R. B. Mazess, in *Noninvasive Measurements of Bone Mass and Their Clinical Application* (S. H. Cohn, ed.), pp. 85–99, C R C, Boca Raton, Florida (1981).
3. L. Burkinshaw, in *Liquid Scintillation Counting*, vol. 5 (M. A. Crook and P. Johnson, eds.), pp. 111–112, Heyden, London (1978).
4. A. Sharafi, D. Pearson, C. B. Oxby, B. Oldroyd, D. W. Krupowicz, K. Brooks, and R. E. Ellis, Multi-element analysis of the human body using neutron activation, *Phys. Med. Biol.* **28**, 203–214 (1983).
5. L. P. Novak, R. E. Hyatt, and J. F. Alexander, Body composition and physiologic function of athletes, *J. Am. Med. Assoc.* **205**, 764–770 (1968).
6. K. Boddy, R. Hume, P. C. King, E. Weyers, and T. Rowan, Total body potassium and erythrocyte potassium and leucocyte ascorbic acid in "ultra-fit" subjects, *Clin. Sci. Mol. Med.* **46**, 449–456 (1974).
7. T. C. Harvey, H. M. James, and D. R. Chettle, Birmingham Medical Research Expeditionary Society 1977 Expedition: effect of a Himalayan trek on whole-body composition, nitrogen and potassium, *Postgrad. Med. J.* **55**, 475–477 (1979).
8. W. E. Siri, in *Advances in Biological and Medical Physics*, vol. 4 (J. H. Lawrence and C. A. Tobias, eds.), pp. 239–280, Academic, New York (1956).
9. W. S. Snyder, M. J. Cook, E. S. Nasset, L. R. Karhausen, G. Parry Howells, and I. H. Tipton, *Report of the Task Group on Reference Man*, Pergamon, Oxford (1975).
10. K. Boddy, *In vivo* activation analysis, in *Progress in Medical Radiation Physics*, vol. 3 (C. G. Orton, ed.) (in preparation).
11. D. B. Morgan, L. Burkinshaw, and C. Davidson, Potassium depletion in heart failure and its relation to long-term treatment with diuretics: a review of the literature, *Postgrad. Med. J.* **54**, 72–79 (1978).
12. G. L. Hill, I. D. McCarthy, J. P. Collins, and A. H. Smith, A new method for the rapid measurement of body composition in critically ill surgical patients, *Br. J. Surg.* **65**, 732–791 (1978).
13. W. B. Nelp, J. D. Denney, R. Murano, G. M. Hinn, J. L. Williams, T. G. Rudd, and H. E. Palmer, Absolute measurement of total body calcium (bone mass) *in vivo*, *J. Lab. Clin. Med.* **79**, 430–438 (1972).
14. A. M. Kodama, N. Pace, and S. J. Parot, Accuracy of measurements of potassium contents of monkeys by *in vivo* body counting as compared to chemical analysis, *Phys. Med. Biol.* **19**, 862–873 (1974).
15. K. G. McNeill, J. R. Mernagh, K. N. Jeejeebhoy, S. L. Wolman, and J. E. Harrison, *In vivo* measurements of body protein based on the determination of nitrogen by prompt γ analysis, *Am. J. Clin. Nutr.* **32**, 1955–1961 (1979).
16. N. Pace and E. N. Rathbun, Studies on body composition. III: The body water and

chemically combined nitrogen content in relation to fat content, *J. Biol. Chem.* **158**, 685–691 (1945).

17. G. B. Forbes, J. Gallup, and J. B. Hursh, Estimation of total body fat from potassium-40 content, *Science* **133**, 101–102 (1961).

18. E. M. Widdowson, in *Human Body Composition: Approaches and Applications* (J. Brožek, ed.), pp. 31–47, Pergamon, Oxford, England (1965).

19. C. T. G. Flear, R. G. Carpenter, and I. Florence, Variability in the water, sodium, potassium and chloride content of human skeletal muscle, *J. Clin. Path.* **18**, 74–81 (1965).

20. J. Bergström, Muscle electrolytes in man determined by neutron activation analysis on needle biopsy specimens. A study on normal subjects, kidney patients, and patients with chronic diarrhoea, *Scand. J. Clin. Lab. Invest. Suppl. 68* **14**, (1962).

21. E. H. Belcher and H. Vetter, *Radioisotopes in Medical Diagnosis*, Butterworths, London (1971).

22. F. D. Moore, K. H. Olesen, J. D. McMurrey, H. V. Parker, M. R. Ball, and C. M. Boyden, *The Body Cell Mass and Its Supporting Environment*, W. B. Saunders, Philadelphia (1963).

23. K. Boddy, P. C. King, R. M. Lindsay, J. Winchester, and A. C. Kennedy, Exchangeable and total body potassium in patients with chronic renal failure, *Br. Med. J.* **1**, 140–142 (1972).

24. F. D. Moore, Energy and the maintenance of the body cell mass, *J. Parent. Ent. Nutr.* **4**, 227–260 (1980).

25. I. Surveyor and D. Hughes, Discrepancies between whole-body potassium content and exchangeable potassium, *J. Lab. Clin. Med.* **71**, 464–472 (1968).

26. B. M. Jasani and C. J. Edmonds, Kinetics of potassium distribution in man using isotope dilution and whole-body counting, *Metabolism* **20**, 1099–1106 (1971).

27. M. Lye and B. Winston, Whole-body potassium and total exchangeable potassium in elderly patients with cardiac failure, *Br. Heart J.* **42**, 568–572 (1979).

28. K. J. Ellis, A. Vaswani, I. Zanzi, and S. H. Cohn, Total-body sodium and chlorine in normal adults, *Metabolism* **25**, 645–654 (1976).

29. S. H. Cohn, The present state of *in vivo* neutron activation analysis in clinical diagnosis and therapy, *At. Energy Rev.* **18**, 599–660 (1980).

30. P. R. J. Burch and F. W. Spiers, Measurement of the γ radiation from the human body, *Nature* **172**, 519–521 (1953).

31. K. Kyere, B. Oldroyd, C. B. Oxby, L. Burkinshaw, R. E. Ellis, and G. L. Hill, The feasibility of measuring total body carbon by counting neutron inelastic scatter gamma rays, *Phys. Med. Biol.* **27**, 605–817 (1982).

32. L. Burkinshaw, I. D. McCarthy, C. B. Oxby, and A. Sharafi, in *Liquid Scintillation Counting*, vol. 5 (M. A. Crook and P. Johnson, eds.), pp. 129–136, Heyden, London (1978).

33. W. E. Siri, Body composition from fluid spaces and density: analysis of methods, *University of California Radiation Laboratory Report 3349* (1956).

34. J. Brožek, F. Grande, J. T. Anderson, and A. Keys, Densitometric analysis of body composition: revision of some quantitative assumptions, *Ann. N.Y. Acad. Sci.* **110**, 113–140 (1963).

35. A. Keys and J. Brožek, Body fat in adult man, *Physiol. Rev.* **33**, 245–325 (1953).

36. A. R. Behnke, E. F. Osserman, and W. C. Welham, Lean body mass. Its clinical significance and estimation from excess fat and total body water determinations, *Arch. Intern. Med.* **91**, 585–601 (1953).

37. J. Mendez and H. C. Lukaski, Variability of body density in ambulatory subjects measured at different days, *Am. J. Clin. Nutr.* **34**, 78–81 (1981).

38. E. Goldman and E. R. Buskirk, Body volume measurement by underwater weighing: description of a method, in *Techniques for Measuring Body Composition* (J. Brožek and A. Henschel, eds.), National Academy of Sciences, National Research Council, Washington, D.C. (1961).

39. J. V. G. A. Durnin and A. Taylor, Replicability of measurements of density of the human body as determined by underwater weighing, *J. Appl. Physiol.* **15**, 142–144 (1960).

40. J. S. Garrow, *Energy Balance and Obesity in Man*, North-Holland, Amsterdam (1974).

41. J. M. Kinney, C. L. Long, F. E. Gump, and J. H. Duke, Tissue composition of weight loss in surgical patients. I: Elective operation, *Ann. Surg.* **168**, 459–474 (1968).

42. T. G. Lohman, Skin folds and body density and their relation to body fatness: a review, *Hum. Biol.* **53**, 181–225 (1981).

43. J. V. G. A. Durnin and J. Womersley, Body fat assessed from total body density and its estimation from skinfold thicknesses: measurements on 481 men and women aged from 16 to 72 years, *Br. J. Nutr.* **32**, 77–97 (1974).

44. L. Burkinshaw, P. R. M. Jones, and D. W. Krupowicz, Observer error in skinfold thickness measurements, *Hum. Biol.* **45**, 273–279 (1973).

45. G. L. Hill, J. A. Bradley, J. P. Collins, I. D. McCarthy, C. B. Oxby, and L. Burkinshaw, Fat-free body mass from skinfold thickness: a close relationship with total body nitrogen, *Br. J. Nutr.* **39**, 403–405 (1978).

46. L. Burkinshaw, The contribution of nuclear activation techniques to medical science, *J. Radioanal. Chem.* **69**, 27–45 (1982).

47. H. Sheng and R. A. Huggins, A review of body composition studies with emphasis on total body water and fat, *Am. J. Clin. Nutr.* **32**, 630–647 (1979).

48. G. L. Hill, J. A. Bradley, R. C. Smith, A. H. Smith, I. D. McCarthy, C. B. Oxby, L. Burkinshaw, and D. B. Morgan, Changes in body weight and body protein with intravenous nutrition, *J. Parent. Ent. Nutr.* **3**, 215–218 (1979).

49. L. Burkinshaw and J. E. Cotes, Body potassium and fat-free mass, *Clin. Sci.* **44**, 621–622 (1973).

50. K. Boddy, P. C. King, J. Womersley, and J. V. G. A. Durnin, Body potassium and fat-free mass, *Clin. Sci.* **44**, 622–625 (1973).

51. G. B. Forbes, Stature and lean body mass, *Am. J. Clin. Nutr.* **27**, 595–602 (1974).

52. J. Womersley, J. V. G. A. Durnin, K. Boddy, and M. Mahaffey, Influence of muscular development, obesity, and age on the fat-free mass of adults, *J. Appl. Physiol.* **41**, 223–229 (1976).

53. F. E. Hytten, K. Taylor, and N. Taggart, Measurement of total body fat in man by absorption of ^{85}Kr, *Clin. Sci.* **31**, 111–119 (1966).

54. G. T. Lesser, S. Deutsch, and J. Markofsky, Use of independent measurement of body fat to evaluate overweight and underweight, *Metabolism* **20**, 792–804 (1971).

55. R. A. Siwek, Multi-element analysis and body composition of the obese subject by *in vivo* neutron activation analysis, Ph.D. dissertation, University of Leeds, Leeds, England (1982).

56. M. A. Smith, in *Noninvasive Bone Measurements: Methodological Problems* (J. Dequeker and C. C. Johnston, Jr., eds.), pp. 77–84, IRL, Oxford, England (1982).

57. S. H. Cohn, D. Vartsky, S. Yasumura, A. Sawitsky, I. Zanzi, A. Vaswani, and K. J. Ellis, Compartmental body composition based on total-body nitrogen, potassium, and calcium, *Am. J. Physiol.* **239**, E524–E530 (1980).

58. R. E. Bigler and H. Q. Woodward, Skeletal distribution of mineralized bone tissue in humans, *Health Phys.* **31**, 213–218 (1976).

59. N. S. J. Kennedy, R. Eastell, C. M. Ferrington, J. D. Simpson, M. A. Smith, J. A. Strong, and P. Tothill, Total-body neutron activation analysis of calcium: calibration and normalisation, *Phys. Med. Biol.* **27**, 697–707 (1982).

60. S. H. Cohn, A. Vaswani, I. Zanzi, J. F. Aloia, M. S. Roginsky, and K. J. Ellis, Changes in body chemical composition with age measured by total-body neutron activation, *Metabolism* **25**, 85–96 1976).

61. S. H. Cohn, in *Noninvasive Bone Measurements: Methodological Problems* (J. Dequeker and C. C. Johnston, Jr., eds.), pp. 17–26, IRL, Oxford, England (1982).

62. A. Horsman, L. Burkinshaw, D. Pearson, C. B. Oxby, and R. M. Milner, Estimating total body calcium from peripheral bone measurements, *Calif. Tissue Internat.* **35**, 135–144 (1983).

63. J. E. Graystone, in *Human Growth* (D. B. Cheek, ed.), pp. 182–197, Lea and Febiger, Philadelphia (1968).

64. D. Picou, P. J. Reeds, A. Jackson, and N. Poulter, The measurement of muscle mass in children using [^{15}N] creatine, *Pediatr. Res.* **10**, 184–188 (1976).

65. R. A. Kreisberg, B. Bowdoin, and C. K. Meador, Measurement of muscle mass in humans by isotopic dilution of creatine-^{14}C, *J. Appl. Physiol.* **28**, 264–267 (1970).

66. G. Greatrex, A. P. Morgan, and F. D. Moore, Evaluation of muscle mass measurement by ^{14}C creatine dilution in the dog, *Metabolism* **21**, 757–760 (1972).

67. A. W. Goode and T. Hawkins, in *Advances in Parenteral Nutrition* (I. D. A. Johnstone, ed.), pp. 557–572, MTP, Lancaster, England (1978).

68. E. C. Anderson, Three-compartment body composition analysis based on potassium and water determinations, *Ann. N.Y. Acad. Sci.* **110**, 189–210 (1963).

69. L. Burkinshaw, G. L. Hill, and D. B. Morgan, in *Nuclear Activation Techniques in the Life Sciences 1978*, pp. 787–798, International Atomic Energy Agency, Vienna (1979).

70. H. C. Lukaski, J. Mendez, E. R. Buskirk, and S. H. Cohn, Relationship between endogenous 3-methylhistidine excretion and body composition, *Am. J. Physiol.* **240**, E302–E307 (1981).

71. R. C. Smith, L. Burkinshaw, and G. L. Hill, Optimal energy and nitrogen intake for gastroenterological patients requiring intravenous nutrition, *Gastroenterology* **82**, 445–452 (1982).

72. F. Ericsson, Intracellular potassium in man. A clinical and methodological study using whole-body counting and muscle biopsy techniques, *Scand. J. Clin. Lab. Invest. Suppl. 163* **42** (1982).

73. J. Mallard, J. M. S. Hutchison, W. A. Edelstein, C. R. Ling, M. A. Foster, and G. Johnson, *In vivo* n.m.r. imaging in medicine: the Aberdeen approach, both physical and biological, *Phil. Trans. R. Soc. Lond. Ser. B.* **289**, 519–533 (1980).

74. W. A. Edelstein, J. M. S. Hutchison, G. Johnson, and T. Redpath, Spin warp NMR imaging and applications to human whole-body imaging, *Phys. Med. Biol.* **25**, 751–756 (1980).

75. J. S. Ingwall, Phosphorus nuclear magnetic resonance spectroscopy of cardiac and skeletal muscles, *Am. J. Physiol.* **242**, H729–H744 (1982).

76. D. G. Gadian and G. K. Radda, NMR studies of tissue metabolism, *Ann. Rev. Biochem.* **50**, 69–83 (1981).

77. R. D. Stoy, K. R. Foster, and H. P. Schwan, Dielectric properties of mammalian tissues from 0.1 to 100 MHz: a summary of recent data, *Phys. Med. Biol.* **27**, 501–513 (1982).

78. L. R. Price, Electrical impedance computed tomography (ICT): a new CT imaging technique, *IEEE Trans. Nucl. Sci.* **NS-26**, 2736–2739 (1979).

79. D. N. Lawley and A. E. Maxwell, *Factor Analysis as a Statistical Method*, 2nd. ed., Butterworths, London (1971).

80. V. D. Barnett, Simultaneous pairwise linear structural relationships, *Biometrics* **25**, 129–142 (1969).

81. D. B. Morgan and L. Burkinshaw, Estimation of nonfat body tissues from measurements of skinfold thickness, total body potassium, and total body nitrogen, *Clin. Sci.* **65**, 407–414 (1983).

3

Medical Applications of Elemental Analysis Using Fluorescence Techniques

PAUL A. FELLER, JAMES G. KEREIAKES, and STEPHEN R. THOMAS

1. INTRODUCTION

Elemental analysis using X-ray fluorescence had its beginnings in the early 1900s with the discovery of characteristic X-ray line spectra. One of the earliest uses of the technique was to determine the elemental composition of mineral samples. During the last 15 years, much progress has been made in X-ray detection and analysis systems, so that fluorescence analysis techniques are being used in many different fields. Modern applications include analyzing environmental pollutants near urban and industrial areas, analyzing commercial products for impurities, determining the constituents of geological and archaeological samples, as well as applications in the field of criminology. In recent years, fluorescence analysis has received much attention from the medical community because of its usefulness in determining concentrations of naturally occurring trace elements and changes in these concentrations due to pathological conditions, and in detecting elements deliberately introduced into the body as tracers.

PAUL A. FELLER • Department of Radiology, The Jewish Hospital, Cincinnati, Ohio 45229. JAMES G. KEREIAKES and STEPHEN R. THOMAS • Department of Radiology, University of Cincinnati College of Medicine, Cincinnati, Ohio 45267.

Elements found in the tissues of animals in very small amounts are called *trace elements*. This term, while not rigorously defined, is generally accepted to mean elements present in concentrations of about 100 parts per million (ppm) or less. The term arose from early reports of investigators who found that several elements were present in tissues in concentrations that at the time were too small to be quantitated. The term *micronutrients* is used to denote trace elements for which essential roles have been demonstrated in warm-blooded animals. If the normal concentrations of these materials are either increased or decreased significantly, interruption of, or interference with, normal life functions will result. At the present time, 15 essential trace elements are recognized in humans. They are Cr, Mn, Fe, Co, Cu, Zn, Se, Mo, Sn, I, V, F, Si, Ni, and As.[1] More information on the roles that trace elements play in the body can be found in the literature.[1-3]

Recent evidence indicates that the concentrations in humans of some essential and nonessential trace elements are different in certain forms of cancer tissue compared to concentrations in normal tissue. Mulay and colleagues[4] examined tissue samples for the presence of 22 trace elements and 2 bulk metals. Concentrations of these 24 elements were measured in cancerous and noncancerous tissue samples taken from the same individuals. Their results showed that for ductal and scirrhous carcinoma of the breast, the Ca, Cu, Mg, Mn, Ti, and Zn content was significantly higher than in noncancerous breast tissue. For bronchogenic carcinoma of the lung, the Zn content was significantly higher and the Fe content significantly lower in the cancerous tissue as compared to the noncancerous bronchial tissue. In adenocarcinoma of the colon, the cancerous tissue had a significantly lower concentration of Sn and a higher concentration of Mn than did noncancerous colon tissue. From their results, it was apparent that concentration differences pertained to a specific type of cancer and that all types of cancer could not be grouped into one category.

Seltzer and colleagues[5] studied serum and tumor Mg concentrations in patients with breast cancer. The cancerous tissues studied contained significantly more Mg than the control breast tissue from the same patient. In the same article, the authors stated that they had also recorded significantly increased tumor Ca concentrations in breast cancer patients.

Danielsen and Steinnes[6] looked at Cu, Zn, K, and Rb concentrations in cancerous and noncancerous human liver samples. Although inconclusive, their tests indicated that the Zn concentration was generally lower and the K concentration generally higher in cancerous tissue. The differences were even more significant if the concentration ratios of K/Zn were examined. The authors felt that the trends they found suggested that further research in this area would be worth while.

Janes and colleagues[7] compared amounts of eight trace metals in normal human bone to that in human osteogenic sarcoma. They found that in the sarcoma samples, Fe, Zn, Cu, and Mn concentrations were increased while those of Ca, Mg, Co, and Ni were decreased.

These investigators and others have found strong evidence that if a sample of tissue is cancerous, it contains certain trace elements in concentrations that are different than those found in normal tissues. Most of the elements cited (10 out of 12) have atomic numbers between 12 and 30. More data from human subjects are needed to make these findings conclusive and to determine if there is a cause and effect relationship between changing concentrations and the diseases. Fluorescence excitation analysis is becoming an increasingly popular technique for determining trace element concentration in human tissues.

In addition to determining concentrations of elements that occur naturally in human tissues, it is also of interest to study the kinetics of elements introduced into the human body in the form of tracers. To date, many of these studies have been performed using radionuclides and radiation detection systems. Now, stable tracers are being used and subsequently analyzed by fluorescence excitation systems. Mention should also be made of fluorescence excitation in diagnostic scanning. However, since this is an imaging modality, it will not be covered in this chapter.

2. PRINCIPLES OF FLUORESCENCE X-RAY EMISSION

2.1. Review of Atomic Structure and Characteristic X Rays

The model of the neutral atom as proposed by Bohr and later more fully explained by quantum mechanics consists of a positively charged nucleus containing Z protons and a varying number of neutrons, surrounded by a negatively charged cloud of Z electrons. It is the atomic number Z that essentially determines the chemical and spectroscopic properties of a group of like atoms and thus defines the element. The phenomenon of fluorescence is produced by the effects of ionizing radiation on the electrons surrounding the nucleus. The orbital electrons are bound to the nucleus by an attractive coulomb force. Quantum mechanical considerations stipulate that these orbital electrons each occupy a specific energy level as defined by a series of quantum numbers. If energy is supplied to an orbital electron in an amount equal to, or in excess of, its binding energy, the electron will be ejected from its subshell, leaving a vacancy or hole. The atom will return to a more stable configuration if this vacancy is filled by a less tightly bound electron from those remaining in the cloud, causing the electron to give up energy equal to the difference in binding energies between the final and initial subshells. If this energy is released immediately (within 10^{-9} sec) after absorption of the incident radiation and is in the form of electromagnetic radiation, the process is called *fluorescence*, and the emitted photon is called a *characteristic X ray*. (If the time period between absorption and emission is more than 10^{-9} sec, the process is called *phosphorescence*). As their name implies, the emitted X rays characterize or identify the element.

2.2. Auger Electrons and Coster–Kronig Transitions

Qualitative identification of an element that has emitted characteristic X rays simply depends on identifying the energy of the radiation emitted. However, quantitatively determining elemental presence is additionally dependent on the rate or intensity of X-ray emission. This is complicated by the fact that two other processes, Auger electron emission and Coster–Kronig transitions, compete with X-ray emission in filling orbital vacancies. In the former, the electron falling into the shell imparts its excess energy to another orbital electron in a lower energy or outer shell, knocking the electron out of the atom's electron cloud. Since this new vacancy is in a less tightly bound shell, the new atomic configuration is more stable than the previous one. However, the atom still has two outer-shell vacancies to fill, one from the original electron transition and one left by the Auger electron. Coster–Kronig transitions are simply low-energy transitions between subshells in one of the major shells and may take place in all but the K-shell. Thus, not all vacancies lead to X rays, and for quantitative analysis, information on the *fluorescence yield*, or the ratio of the number of vacancies resulting in X-ray emission to the total number of vacancies, is needed.

2.3. Matrix Effect

The rate or intensity of characteristic X-ray emission and thus the quantitative determination of elemental presence in a sample is further complicated by the *matrix effect* (matrix refers to the other constituents of the sample containing the elements of interest). The matrix effect consists of relative reductions and/or enhancements of the characteristic X rays of interest due to interactions occurring in the sample itself. These interactions include absorption of the primary excitation radiation before it reaches the element of interest, absorption of the characteristic X rays by the elements of the matrix before they leave the sample, and enhancement of the production rate of the desired characteristic X rays due to secondary and tertiary fluorescence. Secondary fluorescence events in the element of interest are those caused by higher-energy fluorescence radiation emitted by another element in the sample after its excitation by source photons. Tertiary events are those caused by higher-energy fluorescence radiation emitted by another element that has itself been excited by fluorescence X rays from a third element in the sample. Several quantitation methods accounting for the matrix effect in complex samples have been proposed.[8-14] Gardiner and colleagues[15-17] formulated Monte Carlo computer models of gamma ray and X-ray excitation systems to account for or predict the consequences of the matrix effect.

Several authors have published descriptions of the physical principles involved in fluorescence X-ray emission in more detail, and the reader is referred to them for additional information.[1,18,19]

3. DETECTION AND ANALYSIS OF FLUORESCENCE X RAYS

3.1. Overview

Qualitative and quantitative identification of elements present in a sample depends on the ability to distinguish the energies and to count the number of photons emitted from the sample. The original method of separating the spectra made use of diffracting crystals and Bragg's law. By using a crystal with appropriate spacing and a sensitive detector such as a gas-flow counter, collimating the fluorescence X rays emerging from the sample, and varying the incidence angle of the X rays, the characteristic radiation may be analyzed according to wavelength. This type of analysis historically has been called *wavelength dispersive analysis*. This technique has very good wavelength resolution but suffers from low intensity due to the collimation needed to produce a narrow X-ray beam. Another disadvantage is that only one wavelength at a time can be analyzed. Valkovic[1] discusses wavelength dispersive analysis more completely.

In recent years, semiconductor detectors coupled with multichannel analyzers have provided another method for analysis which has been called nondispersive or, more accurately, *energy dispersive analysis*. This technique is now the more common of the two and the method of interest in most medical work. Energy dispersive analysis is usually performed with a solid-state radiation detector and an arrangement of supporting electronics called a pulse processor.[18,20]

3.2. Solid-State Detectors

Semiconductor detectors have provided a great improvement in energy resolution over scintillation detectors and have made multi-element analysis feasible. In this type of detector, an electrical potential or bias is applied across a piece of semiconductor material. Because a high-purity semiconductor is a very poor conductor of electric charge, essentially no current flows across the material. If ionizing radiation hits the material, electrons may be stripped from the atoms of the semiconductor, producing electron-hole pairs. These ion pairs migrate in opposite directions through the material due to the bias voltage and produce a charge pulse that is proportional to the energy deposited by the incident radiation. Measuring these charge pulses yields the energy information necessary to determine the spectrum of the incident beam.

A semiconductor material suitable for use as a radiation detector must possess several characteristics. The *band gap* energy needed to produce an electron–hole pair must be small enough so that the energy lost by the beam will produce a pulse of statistically good size, yet not so small that thermal excitation will produce unwanted *intrinsic* ion pairs. The material must be relatively free of impurities that may either increase intrinsic conductivity or act as traps where the charged ions can be held and thus removed from the pulse. The mobility in the material of the electrons and holes must be good so that collection time is short and recombination is minimized. Also, the material should have a relatively high atomic number so that it is efficient in stopping the incident radiation. The best compromise of these properties to date has been found in germanium and silicon.

It is difficult to produce either Ge or Si of sufficient purity to detect X rays or gamma rays. It takes only small amounts of impurities to combine with the crystal lattice and produce excess free holes or electrons, which cause increased conductivity. To counteract impurities, lithium atoms are drifted into both Si and Ge crystals creating Si(Li) and Ge(Li) material with the desired high resistivity. Even in this configuration, thermal ionization is too high at room temperature to yield suitable results. Thus, these detectors are cooled using liquid nitrogen (boiling point = 77 K). Recently, Ge has been produced in such purity that Li drifting is unnecessary; this material is referred to as *intrinsic* Ge.

Table 1 lists some pertinent properties of Si and Ge. Because Ge has a higher atomic number than Si, it is more efficient at stopping X rays and gamma rays photoelectrically. Thus Ge(Li) or intrinsic Ge is preferred for photons having energies above and about 30 to 40 keV. The fact that more electron-hole pairs are produced at a given energy in Ge implies better energy resolution even at lower energies. However, the larger band gap of Si results

Table 1. Properties of Silicon and Germanium

	Si	Ge
Atomic number	14.0	32.0
Band gap energy (eV)	1.1	0.67
Mean energy expended per electron-hole pair at 77 K (eV)	3.8	3.0

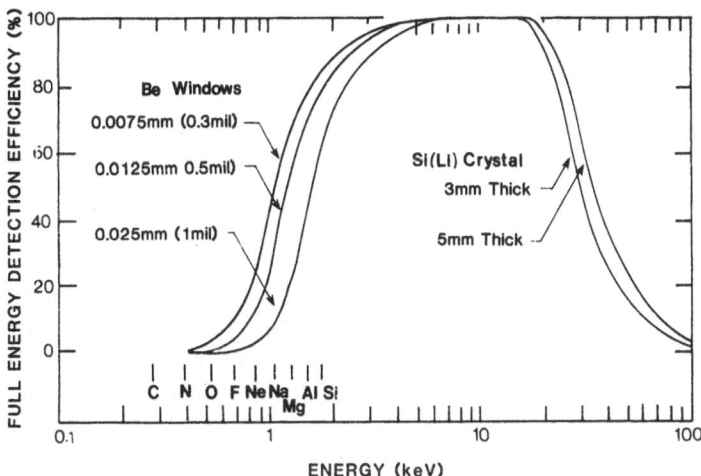

Figure 1. Detection efficiency as a function of photon energy for three Be-window thicknesses and two crystal thicknesses. Elemental symbols indicate the K_α characteristic X-ray energy of that element (see Ref. 21).

in less thermal generation of electrons into the conduction band, and thereby less noise. This makes Si(Li) detectors preferable for low-energy photon detection.

   The cooled semiconductor is kept in a vacuum that provides thermal insulation from the surroundings. A thin Be window in front of the detector provides the separation between the sample and detector environments. Attenuation by this window results in a significant reduction in detection efficiency if the incident X-ray energy is below a couple of keV. However, the window attenuation is actually an advantage when analyzing biological samples for trace elements having $Z \geqslant 12$. This is because 99% of the tissue consists of H ($Z = 1$), C ($Z = 6$), N ($Z = 7$) and O ($Z = 8$), and their very low-energy X rays are more than 99.9% attenuated by the Be window. That means these X rays will not contribute to the dead time of the analyzer and will allow much more rapid analysis of trace elements. Figure 1 indicates the photon detection efficiencies as a function of energy for Si(Li) detectors having various Be window and detector thicknesses.[21]

   As a part of the detector package, the preamplifier stage usually consists of a low-noise field-effect transistor (FET) positioned very near the detector and cooled by the same liquid nitrogen supply. The function of the FET is to integrate the total charge of each pulse coming from the detector and convert it to a voltage signal proportional to the energy deposited in the detector. Since the FET is dc coupled to the detector (low-input capacitance is required to keep electronic noise low), charge builds up as the pulses pass through. A pulsed-optical feedback system is usually employed to reset the charged FET back to zero. That is, when the charge increases to a set limit, a light-emitting diode (LED) is fired and illuminates a photodiode, which then discharges the FET.

Hoffer *et al.*[22] have edited a book describing the principles and medical applications of semiconductor detectors in more detail.

3.3. Pulse Processors

The preamplifier output signal is fed into a main amplifier and associated electronics. The associated electronics usually consist of a pulse shaper, a pileup rejector, a baseline restoration circuit, and a dead-time corrector. These, plus the main amplifier make up the *pulse processor*. In addition, a single or multichannel analyzer (MCA) and a computer or hardwired data processing system may be used.

The main amplifier produces nearly Gaussian output pulses, the amplitudes of which are proportional to the energy absorbed by the detector. For the best noise suppression and energy resolution, the pulse-shaping time constant should be large, yielding wide amplifier output pulses. However, wide pulses present a problem as the input count-rate increases. If a second pulse arrives at the amplifier before the first one has passed through completely, the two pulses will be superimposed, yielding distorted pulses of apparently higher energy. Thus, the selection of the time constant represents a compromise between these two effects.

At higher count-rates, some pulse overlap will occur even with small shaping time constants; this overlap is called *pulse pileup*. A pileup rejector is usually employed to eliminate this undesirable phenomenon. In addition to the main shaping amplifier circuit (commonly referred to as the "slow" amplifier), a second amplifier (the "fast" amplifier) having a very short time constant also receives the preamp output signal. The output of the fast amplifier is a very short pulse for each input or detector event and corresponds in time to the start of a Gaussian signal from the main or slow amplifier. The time intervals between two successive fast pulses is examined. If this interval is such that the two pulses will interfere with each other in the slow channel, then further processing of one or both of the signals is inhibited.

Because it is the amplitude of the slow channel output signal that reflects the energy of the detected radiation, it is important that high count-rates do not introduce dc or baseline shifts and that each Gaussian pulse begins from the proper baseline. This is the function of the baseline restorer.

In quantitative analysis, it is important to correct for dead time, or the time the system is shut down and unable to analyze information from the detector. This is usually done by automatically extending the counting period for the length of time the analyzer is inactive. This correction must take into account dead time during pulsed-optical preamplifier FET reset, during rejection of pileup pulses, and while processing signals in the pulse-height analyzer.

4. EXCITATION MODES

The emission of characteristic X rays from unstable or excited atoms in a sample, the effects of the matrix on the emission of these X rays from the sample, and the detection of these X rays are common to all forms of energy-dispersive fluorescence excitation analysis. The aspect that most distinguishes the various types of fluorescence analysis is the method by which the elements of interest in the sample are excited. Three basic types of excitation will be discussed in the succeeding sections, namely photon, heavy charged particle, and electron excitation.

4.1. Photon Excitation

4.1.1. DESCRIPTION

Photon excitation produces fluorescence X rays through the interaction of incoming electromagnetic radiation with the inner-shell electrons of the sample atoms by the photoelectric effect. The excitation photons may come from a radioactive source or from a system employing an X-ray tube.

4.1.1a. Radionuclide Sources. In radionuclide excitation, the radiation is usually monoenergetic or consists of a few discrete energies. Since the probability of photoelectric excitation of an atom is greatest at incoming energies just above the binding energy of the shell being excited, as seen in Figure 2,[23] it is an advantage if the exciting radiation is limited to these energies. High incoming energies are not only less efficient at excitation but if scattered into the detector will increase instrument dead time and cause more unwanted background counts in the characteristic X-ray region due to Compton scatter in the detector.

The ideal radionuclide source should decay only by emission of one gamma ray of appropriate energy or by electron capture followed by characteristic X-ray emission from the daughter atom. The radionuclide should have a reasonably long half-life for convenience, and its decay product should be either stable or free from interfering emissions. Naturally, such an ideal source does not exist for each element; and since one of the previously mentioned advantages of energy dispersive analysis was its simultaneous multi-element capability, compromises are necessary in select-ing a radionuclide source. A partial list of radionuclides that have proven useful as sources in fluorescence analysis is presented in Table 2. Figure 3

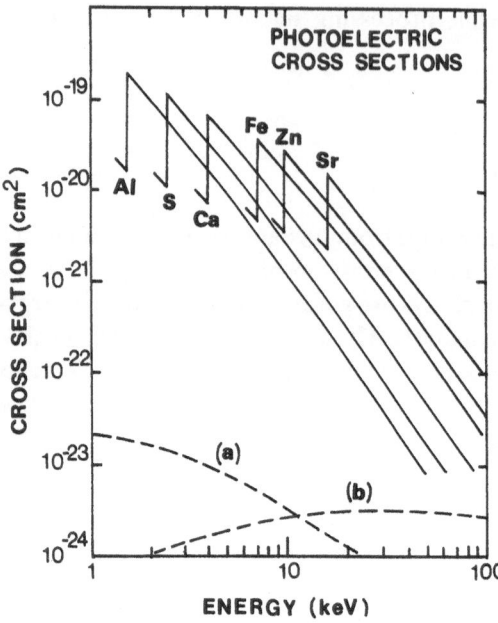

Figure 2. Photoelectric K-shell cross sections as a function of photon energy. (a) Elastic scatter cross section for carbon. (b) Inelastic scatter cross section for carbon (Jaklevic, 1979). Copyright 1979, the Chemical Rubber Company.

Table 2. *Properties of Radionuclides Used as Excitation Sources for Fluorescence Analysis*[a]

Nuclide	Half-life	Emission energies (keV)	Abundance (%)
^{55}Fe	2.7 y	5.8–6.4	26
^{57}Co	270 d	6.4–7.0	51
		14	10
		122	86
		136	10
^{109}Cd	453 d	22–25	102
		88	0
^{125}I	60.2 d	27–32	140
		35	7
^{153}Gd[b]	242 d	41–47	100
		70	3
		97	30
		103	20
^{241}Am	433 y	14–21	33
		60	38

[a] Properties taken from Ref. 24.
[b] Properties taken from Ref. 25.

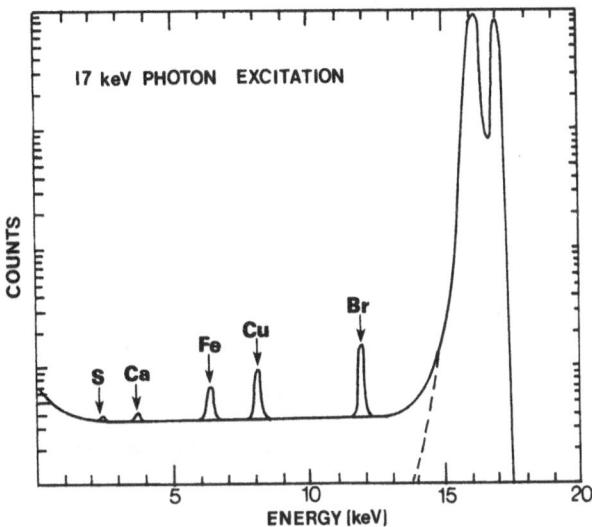

Figure 3. Idealized spectrum representing elements at 10 ppm in a hydrocarbon matrix excited by 17-keV photons. The large peak at 17 keV is due to coherent scattering of the source photons in the sample matrix. The other large, broader peak at just over 16 keV represents incoherent (Compton) scattering from the sample matrix, spread over the possible detection angles (Jaklevic, 1979). Copyright 1979, the Chemical Rubber Company.

presents an idealized pulse-height spectrum for a hydrocarbon matrix containing about 10 ppm of several elements and excited by 17-keV photons.[23]

A major concern with radionuclide excitation sources for trace element analysis is whether sufficient photons are produced from a source of reasonable size or activity. Radiation safety considerations usually require limiting the source activity to under 4 GBq (100 mCi), thus providing photon outputs on the order of 10^7 or 10^8 photons per sec.[26]

The geometry of a radionuclide excitation fluorescence analysis system is a major consideration in order to maximize the excitation and detection efficiencies of the system for fluorescence X rays while minimizing detection of other photons or noise due to scattered radiation and fluorescence from structural materials around the source or sample. With the advent of small solid-state detectors, radionuclide sources are typically in the shape of an annular ring as shown in Figure 4. This type of an arrangement allows the relatively small detector to see a relatively large portion of the sample without being shielded by the source and also provides a large source area, which is useful for radionuclides of low specific activity. In this annular source type of system, the position of the sample relative to the detector and the source annulus has been shown to be very important in terms of X-ray counting rates and peak-to-background ratios.[27,28] Further evaluation of

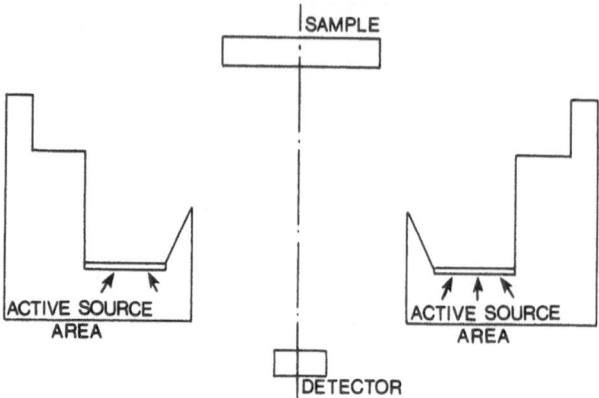

Figure 4. Schematic representation of the system geometry of a radionuclide excitation system with an annular source.

the importance of geometrical factors in this type of system was made by computer simulation and is discussed in a later section. A concentric source sample system has also been developed to reduce relative distances between the source, sample, and detector in order to increase the system's sensitivity.[26]

Radionuclides used as excitation sources in fluorescence anlysis have the advantages of being relatively inexpensive and convenient to use. Radioactive decay provides a predictably constant source of photons. However, there are limitations to the activity levels of these sources that can be conveniently and safely handled, and thus the photon flux available to excite the sample is generally less than that achieved with sources that produce photon beams electronically.

4.1.1b. X-Ray Tube Sources. In contrast to radionuclide sources, X-ray tubes usually provide a greater photon output. However, this output is in the form of a bremsstrahlung continuum spectrum. This means a trade-off of high photon yield versus relatively high system dead time and increased background in the region of interest, especially for low atomic number elements or for trace elements in a low-Z matrix. Figure 5 represents a typical spectrum for the same sample in Figure 3 but excited by a continuous X-ray spectrum of energies.[23] One compromise is having the beam from the X-ray tube excite a secondary fluorescer target whose characteristic X-ray energies are just above the binding energy of the element of interest. This secondary target then acts as a source of photons to excite the sample. Since the efficiency of characteristic X-ray production in the secondary fluorescer target is much less than 100%, there is a significant decrease in the number of source photons striking the sample with this method. Another technique used to provide a more nearly monoenergetic photon beam from an X-ray tube

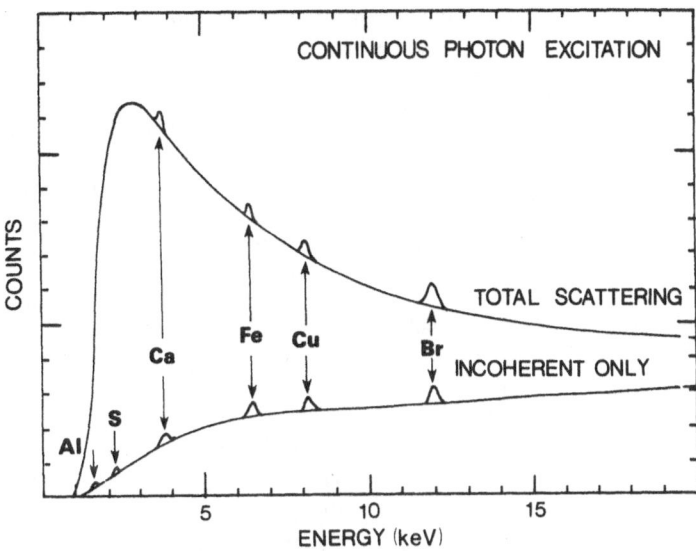

Figure 5. Idealized spectrum representing elements at 10 ppm in a hydrocarbon matrix excited by a continuous photon spectrum. The two curves represent the two extreme cases of probability for source photons scattering from the sample matrix. The upper curve reflects a high probability of coherent scattering, and the lower curve reflects a zero probability of coherent scattering (Jaklevic, 1979). Copyright 1979, the Chemical Rubber Company.

involves using regenerative monochromatic filters (RMFs). These filters have high transmission just below the K absorption edge but low transmission above and below that energy. In addition, using different anode materials in the X-ray tube can result in different spectral distributions of source energies.

Gedcke and associates[29] have made an intercomparison of trace element excitation methods using X-ray tubes with the following commercially available configurations: (1) broadband excitation using the primary anode, (2) RMFs using various X-ray tube anode materials, and (3) secondary fluorescers. Their aim was to evaluate general-purpose excitation techniques for effectiveness in trace element detection in a low atomic number matrix. The test sample used was a $Na_2B_4O_7$ fused glass standard containing trace amounts of six elements. Table 3 lists their resulting values of minimum detectable concentrations obtained from four different commercially available instruments. Values in the table are corrected for differences in spectrometer performance. Their conclusions were that 15-kVp broadband excitation is optimum for light-element excitation, while monochromatic excitation is best for the 5–30-keV energy range. The 50-kVp broadband excitation was superior to monochromatic excitation if simultaneous analysis of a wide range of elements was desired.

Table 3. Minimum Detectable Concentrations (MDC) of Trace Elements in a Glass Standard[a]

Trace element[e]	Secondary fluorescer				R.m.f.		Broadband 15 kVp	
		MDC[b](ppm)			MDC[b](ppm)		MDC[b](ppm)	
	Fluorescer	A[c]	B[c]	Anode/filter	C[c]	D[c]	C[c]	D[c]
Na K$_\alpha$	Cl	570[d]					410	300
	Ti	1100	3400[d]					
Al K$_\alpha$	Cl	41[d]					24	27
	Ti	36	90[d]					
K K$_\alpha$	Ti	1.3	4.7[d]					
	Cu	4.4		W/Cu	9.9[d]	9.6	7.5	4.7
Ba L$_\alpha$	Cu	4.8		W/Cu	6.3[d]	5.6	10	12
C K$_\alpha$	Cu	0.64		W/Cu		1.4		
	Mo	2.8	2.2	Mo/Mo	3.4	1.9		2.7
Ge K$_\alpha$	Mo	0.29	0.24	Mo/Mo	0.25	0.18		

[a] Adapted from Ref. 29.
[b] Corrected for spectrometer differences.
[c] A, B, C, D represent the commercially available analysis systems used.
[d] Limited by maximum tube current.
[e] Shown with characteristic X ray.

Ong and colleagues have reported the development of two spectrometers that achieve monochromaticity by using a secondary fluorescer,[30,31] and a line-focusing cylindrically curved Johanson-bent crystal.[31,32] They report that the latter system results in a cleaner excitation spectrum and can also provide a focused beam of high power density.

Several authors [33-37] have discussed the use of polarized incident radiation to reduce the scatter, and thereby increase the signal-to-noise ratio in the analysis spectrum. The relatively high background observed with X-ray tube excitation is due to scattering of the primary beam in the sample. By arranging the X-ray source, sample, detector, and a scattering or polarizing material along three mutually perpendicular axes as in Figure 6, it is possible to minimize the amount of scattered radiation reaching the detector. This is due to the fact that the scattering intensity is zero in the direction of the incident radiation's electric vector. Radiation emerging from an X-ray tube is unpolarized. That is, electric vectors may have any orientation in the plane perpendicular to the direction of propagation of the radiation. However, all scattered radiation emerging from the polarizer at 90° to the incident beam will have its electric vectors pointing perpendicular to the plane of scattering. When this radiation is incident on the sample, there will be no scattering in the direction of the electric vector. Thus, a detector placed along that axis will

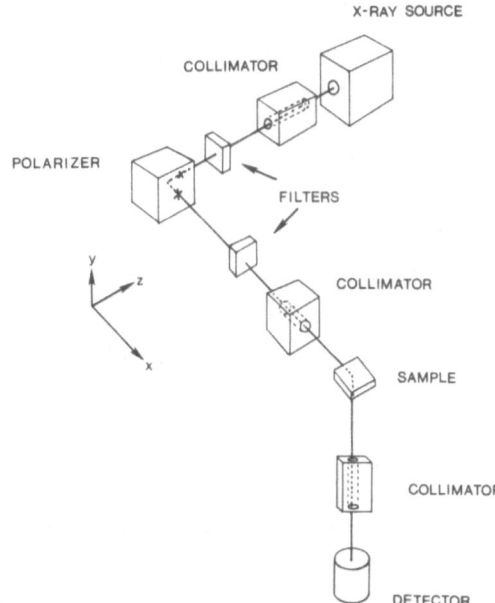

Figure 6. Representative three-axis geometry of an X-ray fluorescence system using polarized incident radiation.

detect minimal scattered radiation. Since fluorescence radiation is emitted essentially isotropically, its intensity is unaffected. This technique is limited by the severe reduction in intensity of the exiciting radiation but has been shown to be useful in analyzing thick biological specimens that produce a relatively intense scatter spectrum. Ryon[37] has suggested, and Standzenieks and Selin[38] have tested the use of this three-axes geometry, but they have substituted secondary fluorescers for the polarizer. This will reduce that portion of the background spectrum which is due to scatter of the primary radiation in the secondary fluorescer.

As previously mentioned, the advantage of greater photon output from the X-ray tube must be weighed against the possibility of increased background in the energy region of interest and higher system dead time due to the wide range of source energies. The use of secondary fluorescers, and the ability to change the output spectrum by selecting proper X-ray tube accelerating potentials, beam filters, and anode target materials, make this form of excitation very versatile. Either method of photon excitation offers the additional advantage of sources that are relatively inexpensive and easily placed in a clinical setting. Generally, mixed-energy sources (such as X-ray tubes) are less desirable than monoenergetic sources for *in vivo* excitation. This is because all primary or source photons can contribute to the radiation dose to the patient, but only those just above the absorption edge energy are highly efficient for fluorescence X-ray production.[39]

4.1.2. TARGET PREPARATION

One of the strengths of X-ray fluorescence analysis using photon excitation is that samples having many different forms may be analyzed. Sample targets may be in the form of solids, powders, pellets, liquid solutions, and even gases. Due to the diffuse nature of source emissions, photon excitation is best suited for relatively large (1–10 cm^2) sample cross-sectional areas. Using thin samples reduces absorption and scatter effects. Homogeneous samples produce more consistent results and reduce the complexity of corrections that may be needed for material and particle-size effects.

Solid samples, such as pieces of bone, teeth, or hair, may require little or no sample preparation. When uniform tissue samples are desired, or when a preconcentration of the elements of interest is needed in tissue or blood samples, samples may be ashed and used in powder form or pressed into pellets. Care must be taken in these procedures so that new elements or changes in the relative concentration of elements already present are not introduced. In some cases, liquid samples can be analyzed with little or no sample preparation. If the liquid is a true solution, it will be perfectly homogeneous. Internal standards can be added and samples can be concentrated or diluted easily. One technique of preconcentrating elements of interest if they are suspended in a liquid sample is to evaporate the liquid and analyze the residue, or pass the suspension through millipore filters and analyze the resulting deposits.

4.1.3. MEDICAL APPLICATIONS

During the last decade, many medically related fluorescence studies using photon excitation have been reported. They have included analyses for (1) naturally occurring elements in tissues and blood; (2) elements associated with a pathological condition or accidently introduced matter (foreign particles, toxic substances); and (3) deliberately introduced foreign material (tracers, labels, drugs, etc.). The following represents only a partial list of these reports.

4.1.3a. Radionuclide Studies. In studying the effect of osteoporosis, osteomalacia, and rheumatoid arthritis, it is desirable to have information on the mineral density of bone. Patomäki and Olkkonen[40] have described a method for determining the mineral density of trabecular bone *in vitro* by measuring the Ca content of bone with an X-ray fluorescence analysis system employing 740 MBq (20 mCi) of ^{55}Fe. They found a linear correlation between the bone mineral density and the mean Ca X-ray count rate and used this linear relationship to calibrate their system. They also developed a method for measuring structural inhomogeneity by calculating the coefficient of variation from ten measurements of Ca X-ray count-rate taken along a line at 1-mm intervals. The more inhomogeneous the bone sample, the greater the coefficient of variation. Their studies on this system have shown a decrease in trabecular bone mineral density with patient age. They also noted a decrease in bone mineral density and an increase in the coefficient of variation in osteoporotic samples as compared to normal samples. The authors hope that their studies will give more information about the suitability of the X-ray fluorescence method for clinically diagnosing bone diseases.

Cesareo and Del Principe[41] determined the iron content in dried venous blood on filter paper by exposing samples to a 370-MBq (10 mCi) annular source of ^{238}Pu and analyzing the iron X rays using a proportional gas counter. The authors report that the technique, which requires minimal sample preparation and a small quantity of blood, permits rapid analysis of K and Ca as well as Fe and that drying the blood causes an Fe preconcentration in the sample. They report a sensitivity limit of about 10 ppm Fe in one thousand sec.

Ahlgren and Mattsson[42] reported an X-ray fluorescence technique for *in vivo* determination of Pb concentration in bone using a 740-MBq (20 mCi) ^{57}Co source. The minimum detectable Pb concentration in a finger bone was 14 µg/g for a 15 min measuring time. They used the method to measure Pb concentration in metal industry workers.

The same investigators[43] measured Cd in the kidney cortex of human subjects *in vivo* by fluorescence excitation analysis employing an 11-GBq (300 mCi) ^{241}Am source. The authors report that the minimum detectable Cd concentration varies between 20 and 40 µg/g $\pm 30\%$ for distances between the skin and the kidney surface of 30 to 40 mm. They suggest that this technique could be used as a substitute for neutron activation analysis to measure the body's burden of Cd in occupationally exposed individuals.

In vitro X-ray fluorescence analysis of stable tracers is useful for measuring various body compartments or extra cellular fluid spaces in humans. This is analogous to using radioactive tracers that are introduced into the body and then analyzed *in vitro* in small samples of tissue or body fluids extracted from the patient. The advantage of the fluorescence analysis procedure is the absence of radiation exposure to the patient. Kaufman, Price, and colleagues[44,45] have been very active in this area. They described the determination of the extra cellular fluid volume in humans using orally administered stable sodium bromide as the tracer then subsequently analyzing plasma samples for stable Br content by fluorescence analysis employing a 92.5-MBq (2.5 mCi) ^{109}Cd source. They compared their results from this technique to results obtained from the more established method using radioactive ^{82}Br and reported very good agreement. This group later reported the development of an automated fluorescence excitation analysis system for medical applications that has a 22-GBq (600 mCi) ^{241}Am source and a 2.2-GBq (60 mCi) ^{109}Cd source available for analysis.[46] The researchers also developed a technique for determining red cell volume in humans that uses nonradioactive Cs as an analogue of K for labeling red blood cells.[47] Their technique consists of assaying the initial and equilibration concentrations of Cs by fluorescence excitation analysis using the 22-GBq [241]Am excitation source. In their comparative study in 11 humans, the resulting red cell volume calculated by the fluorescence excitation analysis technique was in good agreement with that calculated by the traditional radioactive ^{51}Cr method.

There has been an interest for many years in the kinetics of the various radio-opaque contrast media that are used in radiological studies. Moss *et al.*[48] discussed a fluorescence excitation technique using an 18.5-GBq (500 mCi) [241]Am source for the *in vitro* measurement of nonradioactive I concentration in tissue samples. Similar techniques were found to be useful in studying contrast media,[49] although much of the work to date has been on animals rather than humans. Kaufman and colleagues[50] have suggested an absorption correction technique that may extend the applicability of fluorescence analysis to quantitating contrast media *in vivo*.

In clinical medicine, it has been common to use plasma creatinine levels and blood urea nitrogen (BUN) to indicate renal function. However, these indicators provide only rough estimates of kidney function and may not be abnormal until the glomerular filtration rate (GFR) is already abnormally low. Alazraki and colleagues[51] have described a noninvasive method of determining GFRs by infusing small amounts of nonradioactive iothalamate and subsequently collecting plasma samples to be assayed by X-ray fluorescence. Their experimental technique employed two 37-GBq (1 Ci) sources of [241]Am to excite the stable I in the iothalamate in the collected plasma samples. Their animal data demonstrated good correlation with that of a classical ^{14}C-inulin procedure. Guesry and associates[52] have compared fluorescence analysis of iothalamate with both the inulin clearance and ^{125}I-iothalamate techniques and have found it to be a valid estimator of GFR.

Cesareo and colleagues[53] have suggested that labeling such blood components as erythrocytes, leukocytes, platelets, etc., with stable, rather than radioactive tracers and their subsequent analysis by X-ray fluorescence techniques could be used for determining the life span of these components and for studying their metabolism. They labeled platelets from human blood samples with stable selenocystine, deposited the platelets on millipore filters and analyzed them for Se content using the [238]Pu-excited system mentioned earlier. *In vitro* survival curves are reported to be in accord with values reported in the literature.

In another paper, Cesareo[54] reported animal studies where the cardiac output was measured by injecting nonradioactive I intravenously and determining its concentration in arterial blood samples deposited on filter paper by measuring the characteristic X rays emitted by the I when excited by a [238]Pu source. The authors state that their ultimate goal is to measure externally the cardiac output without taking blood samples by directly placing the counter and the radioisotope source over the thyroid area. This would allow the simultaneous measurement of I absorption by the thyroid.

The application of radionuclide-excited fluorescence excitation analysis to the study of cerebral blood flow has been suggested by Hoffer, Moody, and colleagues.[55, 56] Their experimental apparatus employed a 122-GBq (3.3 Ci) [241]Am annular source. One *in vivo* monkey experiment incorporated an intra-arterial injection of a tracer containing stable I and subsequent *in vivo* analysis by placing the apparatus over the animal's head. In a second experiment, a monkey was forced to breathe a mixture of stable Xe gas and O_2 for 6 min, after which the system was abruptly flushed by 100% O_2. The analysis system was placed over the head and used to determine Xe washout from the region of interest in the white matter of the brain. The feasibility of this technique was tested in a couple of human patients.[56] The feasibility of the I tracer method was limited by the undesirability of an intra-arterial injection but was felt to be useful in patients undergoing routine arteriography. The authors felt that the technique employing Xe gas could have wider applications to humans than the I technique.

The suitability of radionuclide ([55]Fe) excitation in determining the concentration of Ti, Ca, and Mg in biological samples has been investigated.[28] Evaluations in pellets of simulated biological samples of very low atomic number ($Z < 12$) doped with various concentrations of the elements of interest were carried out through Monte Carlo computer simulation by modifying the software of Gardner and colleagues.[15] The Monte Carlo simulations were validated by varying the configuration for a radionuclide excitation system employing 3.7 GBq (100 mCi) of [55]Fe shown schematically in Figure 4 and comparing values of relative response obtained experimentally with those predicted by the computer.

Figure 7 indicates the experimental and computer-predicted relative system responses to Ti versus sample position. The importance of sample position is evident and compatible with experimental results of Vāñó and González.[27] Figure 7 also indicates close agreement between the experimental and computer-predicted values. When this agreement between computer and experiment was obtained, modifications in collimator height, source-to-sample and detector-to-sample distances, detector radius and inner and outer radii of the source annulus were made in the computer simulation, and predicted changes in system sensitivity were obtained.

Table 4 lists qualitatively various changes in geometrical constants made in the computer simulation and the corresponding advantage factors (AF) or new response rates relative to the original response rates for each optimization change. These results indicate that through judicious selection of these various factors, the sensitivity of such an analysis system could be increased by nearly two orders of magnitude for Ti and Ca and by about 30 times for

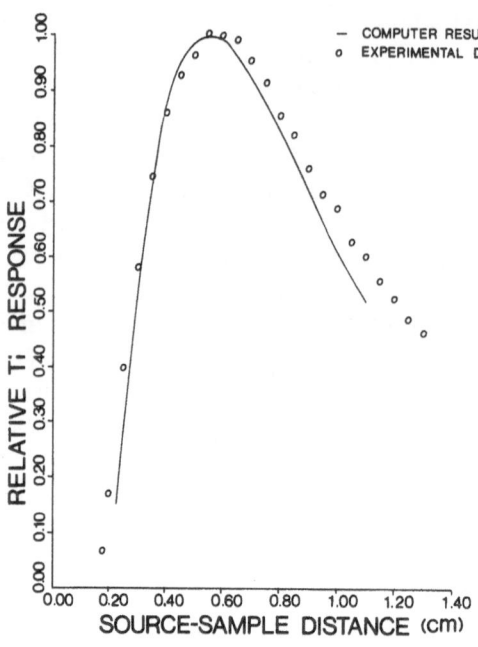

Figure 7. Comparison of experimental and Monte Carlo-predicted Ti responses of an annular ^{55}Fe-excited fluorescence system as a function of sample position.

Mg. It can be shown that the change in the minimum detectable concentrations is proportional to $1/\sqrt{AF}$. These increases in sensitivity would thus correspond to a reduction in the minimum detectable concentration of Ti and Ca by a factor of about 10 and for Mg by a factor of about 5. Figure 8 shows the computer-predicted minimum detectable concentrations as a function of the atomic number for such an optimized system employing an ^{55}Fe excitation source. With such an optimized system, both Ti and Ca analyses would be possible at concentration levels reported in tissues.[4] However, Mg analysis would not be possible at the levels reported.

4.1.3b. X-Ray Tube Studies. Using whole blood as a representative sample, X-ray fluorescence employing X-ray tube excitation was evaluated by

Table 4. *Computer Optimization Adjustments and Corresponding Advantage Factors for a Fluorescence Analysis System Employing an* ^{55}*Fe Excitation Source*

	Ti	Ca	Mg
Reduce collimator height	1.41	1.52	2.61
Move detector nearer sample	6.13	6.50	3.98
Increase detector radius	7.29	6.98	2.35
Reduce source radii	1.36	1.37	1.27
Total advantage factor	86	95	31

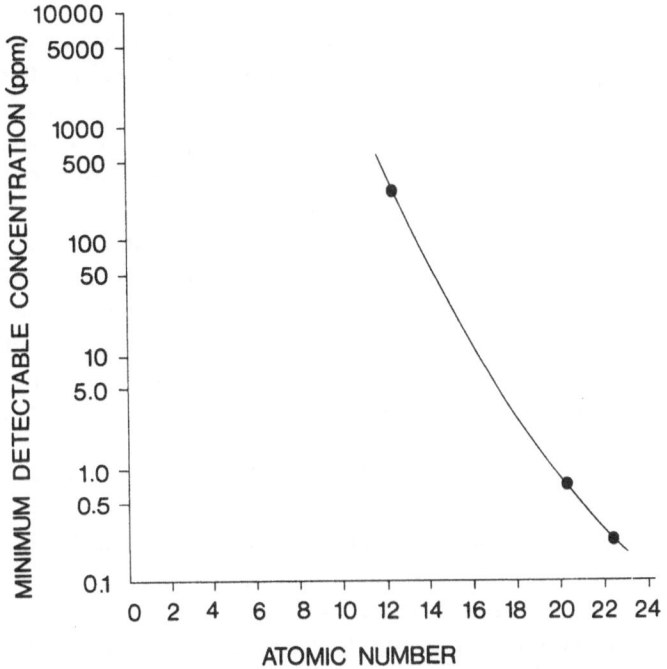

Figure 8. Monte Carlo predictions of minimum detectable concentrations for Mg, Ca, and Ti with a computer-optimized ^{55}Fe-excited fluoresence system.

Lubozynski et al.[57] Outdated whole blood samples were doped with liquid solutions of the elements of interest and fluorescence measurements performed on lyophilized samples. The authors tested the detectability of eight elements (Cr, As, Sr, Ag, Cs, W, Hg, Pb) and used X-ray tubes with either a W or Mo anode. They reported lower limits of detectability ranging from about 2 to 15 ppm for elements whose characteristic X rays fell between the ranges 4–6 keV or 9–14 keV. They also tested relative sensitivities for one element in blood when in the form of liquid, dry powder, or dry powder pressed into pellets. The dry powder and the pellets yielded sensitivities that were 3 times and 3.3 times, respectively, the sensitivity of the liquid.

Ong[31] reports the use of his secondary fluorescer excitation system[30] for the analysis of five major elements in serum for diseased versus normal patients and reports interesting changes in Cu and Zn levels. However, due to his limited number of observations, he does not attempt to draw any conclusions from the data. He has done additional studies of trace elements in human hair with this system and suggests that fluorescence techniques using this system may prove useful in studying the role of metalloproteins in biology.[31] Cox and Ong[58] reported a method for determining the mass of soft-tissue samples using the Compton-scatter intensities observed in their analyzers.[30,32] This method reportedly enables elemental weight-fraction concentrations to be measured with a high degree of precision and accuracy.

Jaklevic and associates[59] describe a photon-excited X-ray fluorescence system optimized for the accurate and sensitive analysis of small (1 mm² or less) areas. The unit uses an automatic scanning device that advances the sample past the analysis region and was designed for measuring trace element profiles along the length of individual human hairs.

A fluorescence method of *in vivo* cardiac output determination has been described by Kaufman *et al.*[60] The source-detector system consisted of a 95-kVp, three-phase X-ray generator and a Si(Li) detector oriented perpendicularly to the source photon beam. Their tests, performed on dogs, were done by positioning the source-detector system so that the sensitive analysis region was a small volume of blood in the heart, then analyzing the I content after a bolus intravenous injection of stable iothalamate contrast material. The authors obtained results that were in reasonable agreement with other methods. A later publication[50] described an absorption correction technique that reportedly improves the accuracy of *in vivo* I-concentration measurements.

Phelps and colleagues[61] reported the *in vivo* measurement of regional cerebral blood volume by X-ray fluorescence of an ingested iodinated contrast material. Their technique used an X-ray tube operated at 80 kVp and filtered with 1 mm Cu. When tested in dogs and monkeys, the results of the *in vivo* technique agreed within 8% with results of an *in vitro* ^{51}Cr-labeled red blood cell method. The researchers predict that the error in human subjects may be greater due to the low transmission of the I X rays through the human skull.

4.2. Particle-Induced X-Ray Emission (PIXE)

4.2.1. DESCRIPTION

In the particle-induced X-ray emission technique, charged particles are accelerated to an appropriate energy and finally collimated and focused on the sample to be analyzed. The X rays, characteristic of the elements present in the sample, are produced during the bombardment and detected by a semiconductor detector. The types of particles used may be classified as follows:

4.2.1a. Protons and Alphas. Although using charged-particle accelerators permits using a wide variety of projectile types and energies, developments in the field have centered on two principal methods typically using 3-MeV protons and 20-MeV alpha particles. The energies are chosen so that the background continuum distribution does not interfere strongly with

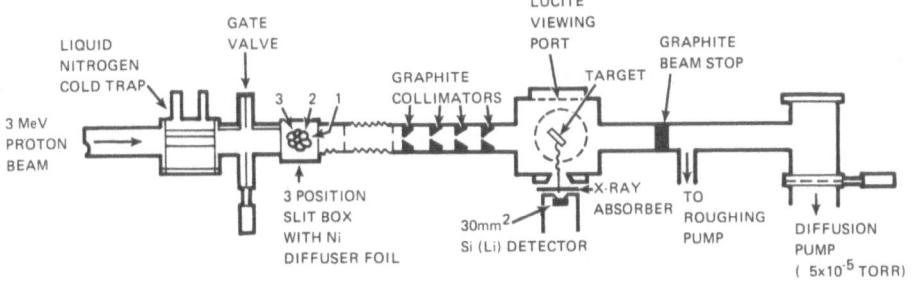

Figure 9. Schematic of a typical charged-particle fluoresence analysis system (Jaklevic, 1979). Copyright 1979, the Chemical Rubber Company.

the characteristic X-ray region of interest. Figure 9 shows a diagram of a typical charged-particle excitation apparatus.[23] The high-energy beam is directed down an evacuated beam line toward the target. By adjusting a magnetic focusing element either a focused or a diffused beam can be obtained. Samples are normally contained in multiple-target holders to reduce the frequency of entry into the vacuum chamber. The beam passes through the thin sample and is absorbed by the graphite beam stop that is shielded from the detector's field of view. The total current can be carefully measured and results used to calibrate the system. The X rays are detected through a thin beryllium window inside the chamber.

Figure 10 shows the cross sections for K-shell ionization for a few elements as a function of proton energy.[23] The cross sections for these elements have maximum values that are roughly an order of magnitude lower than values for photoelectric ionization, and they exhibit a more gradual energy dependence (see Figure 2). It is seen from Figure 10 that the cross section (and, accordingly, the sensitivity) of a given element is maximum at a specific proton energy, which, unfortunately, differs for different elements. For this reason, it is impossible to achieve optimum sensitivity for all elements in one bombardment run. In practice, this is not anticipated as a serious restriction and can be circumvented when necessary by carrying out sequential bombardments at different projectile energies. Similar results can be calculated for alpha particles. Figure 11 depicts an idealized pulse-height spectrum for the same sample in Figure 3 but for proton excitation. The continuum background is created mostly by the energetic electrons ejected in the ionization process rather than by the incident particles themselves. This second-order process increases the minimum detectable limits for proton excitation to about the levels obtainable by photon excitation.

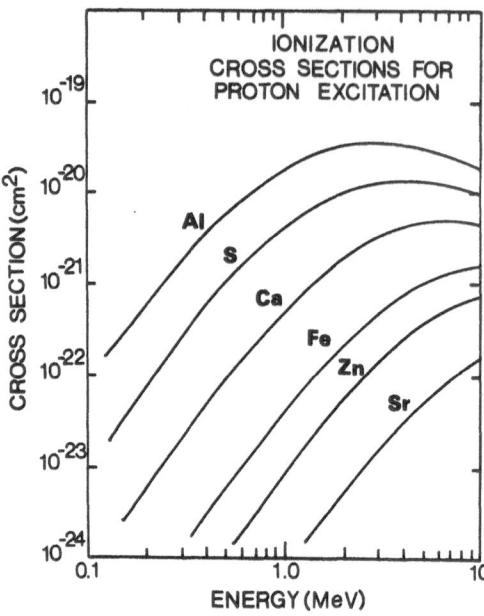

Figure 10. K-shell vacancy production cross sections as a function of proton energy (Jaklevic, 1979). Copyright 1979, the Chemical Rubber Company.

4.2.1b. Heavy Ions. In general, maximum cross sections are proportional to the square of the charge of the projectile and occur at energies directly proportional to the mass of the projectile. From this information, it can be concluded that best results will be obtained using a completely stripped heavy ion at the appropriate energy. Folkman[62] has discussed the processes involved in generating X-ray spectra by bombarding different materials with heavy ions and resulting implications for analysis. Impetus for collision studies with heavier ions was gained after the discovery of regular fluctuations in K, L, and M X-ray production cross sections as a function of target atomic number when a variety of target elements were bombarded with fission products. These variations showed enhanced X-ray production whenever the target and projectile energy levels matched. Cross sections for X-ray production by heavy ions are several orders of magnitude higher than those for X-ray production by protons of the same energy. Since heavy ions penetrate only a few hundred Å (less than 100 nm) into most materials, the technique can be used as the basis for examining surface phenomena.[1]

4.2.1c. Muons. Negative muons are produced by the in-flight decay of negative pions, which are artifically produced by bombarding a target with photons or electrons having kinetic energies greater than the rest mass of the pion. The muon has a mean life that is nearly two orders of magnitude larger than the pion (2.2×10^{-6} and 2.6×10^{-8} sec, respectively), while their rest masses are similar (105.7 and 139.6 MeV, respectively). The free muon decays

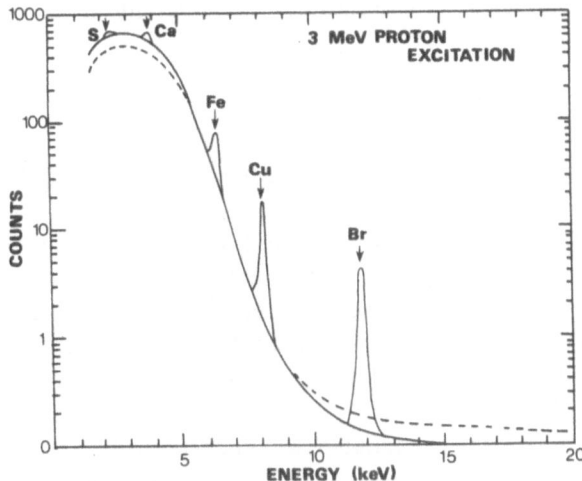

Figure 11. Idealized spectrum representing elements at 10 ppm in a hydrocarbon matrix excited by 3-MeV protons (Jaklevic, 1979). Copyright 1979, the Chemical Rubber Company.

into an electron and two neutrinos. The sequence of events taking place in the so-called muon channel in the accelerator is shown in Figure 12.[1]

In passing through matter, muons, like other charged particles, are slowed down by ionization processes. When a negatively charged muon comes to rest, it is captured by a nearby nucleus to form a muonic atom, analogous to an electronic atom. Since all of the muonic atomic levels are unoccupied, the muon will cascade down through these levels, and in the process, the atom will emit X rays characteristic of the muonic energy levels of the capturing nucleus. These X rays are some 200 times more energetic than ordinary atomic X rays due to the high rest mass of the muon, which is about

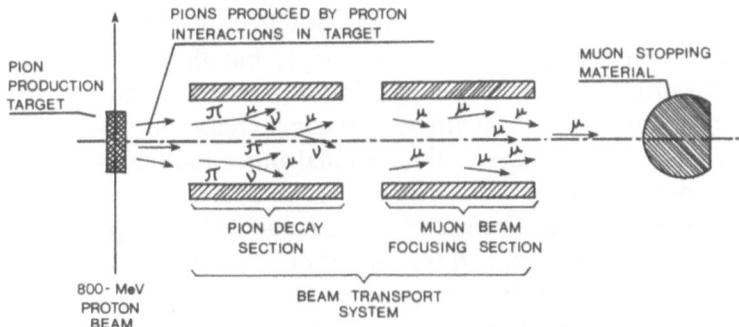

Figure 12. Schematic representation of the muon channel of a charged particle accelerator (Valkovic, 1980). Copyright 1980, the Chemical Rubber Company.

207 times that of the electron. Valkovic[1] discusses muonic X-ray production in greater detail.

4.2.2. TARGET PREPARATION

Target preparation is one of the central problems in X-ray emission spectroscopy analysis. Sample excitations by proton (or any other charged-particle beams) can be accomplished with thin targets, although the simplest method is the direct bombardment of a thick specimen. This method is often used for biological materials, such as teeth, from which it is difficult or impossible to obtain thin targets. Drawbacks of this method are mainly the lower sensitivity and the need for various corrections in calculating the results.

When thin targets are prepared, the material to be analyzed should be deposited on some suitable backing. The backing material should have good mechanical strength, good electrical and thermal conductivity, and high purity, and should be able to withstand high beam intensities. The continuous background radiation produced by the backing should be as small as possible, which favors thin backings consisting of low Z elements, such as thin Ca or plastic foils.

Target preparation for irradiating by charged particles often requires the samples to be solubilized. For most biological substances, the two common procedures for organic matter destruction are wet digestion and dry ashing. Several workers have made thin targets by depositing a very small mass (less than 1 mg) of lyophilized or ashed material on a backing foil, with adhesion affected by adding a drop of wetting agent or glue. Water or some other liquid samples can be prepared by drying a few drops on an appropriate backing when charged particles are used for excitation.[1]

Separating elements of interest from a large volume into the almost ideal matrix and volume of a resin-loaded paper disc has proven to be a versatile and effective approach to trace metal determination. The chemical preconcentration not only reduces or eliminates problems of matrix correction and variations in physical properties of the sample, but also increases sensitivity by several orders of magnitude. In addition, reliable standards are easily prepared, usually from standard reference materials with compositions or physical properties certified by the National Bureau of Standards.

In order to achieve optimal sensitivity in analysis, thin substrates are used whenever possible. For example, the range of 3-MeV protons in Ca is approximately 16 mg/cm^2, and the ionization process is most effective over the early part of the energy loss process. Thus, the highest accuracies and sensitivities are achieved with samples having thicknesses of less than 1 mg/cm^2. However, the thin substrate limits the amount of beam current that can be directed on the sample, since local heating in the nonconducting vacuum surroundings can evaporate more volatile constituents or physically damage the sample. Using conductive substrates, such as thin Ca foils, reduces the problem and also eliminates charging effects on the sample. For quantitative analysis of thicker samples, the effects of energy loss of the incident protons and absorption of fluorescence X rays must be considered. For thick biological specimens, electrostatic charging of, and possible evaporation of, volatile constituents from the sample must still be considered.[23]

4.2.3. MEDICAL APPLICATIONS

The presence of traces of heavy elements in the human organism is of great interest in detecting and probably treating human diseases. Particle-induced X-ray emission analysis has already become a very effective means of detecting trace elements in biomedical samples, and sensitivities from 0.1 to 1.0-ppm wet weight can now easily be obtained. These sensitivities can generally be achieved in approximately 15 min, using as little as 100 μg of material. If necessary, an analysis can also be easily performed on biopsy tissues, using even less material. Another advantage of PIXE analysis stems from the fact that 10–15 elements can simultaneously be detected in a human organ in a single analysis.

Of particular importance is establishing normal concentrations of trace elements in organs and tissues of "healthy" individuals. A very useful compilation on the elemental concentrations in human organs, in both normal and diseased states, is given by Anspaugh et al.[63] Trace element concentrations in different organs have been studied for some time. The accumulation of certain metals in some organs in malignant diseases has attracted special attention.

Liver tissue has been studied by several research groups. Dabek et al.[64] have used the technique of proton-induced X-ray analysis for studying Cu, Zn, and Fe in human liver tissue in samples obtained postmortem from normal and diseased (cirrhotic) cases. Cirrhotic tissue showed a substantial rise in the Cu concentration and, in some cases, a rise in Zn concentration.

Kubo[65] made measurements on three sets of human liver samples to determine the precision of proton-induced X-ray fluorescence analysis. Two specimens were obtained from an individual who died of acute myelocytic leukemia. Two additional specimens were taken from tissue sections of the same liver samples but containing metastatic foci. Within each set, concentrations of trace elements were reproduced to within 20%. In some later investigations, Kubo et al.[66] used proton-induced X-ray fluoresence to determine the concentration of trace elements in human prostate and liver tissue samples; Ca, Fe, Cu, and Zn were observed in all specimens. Small amounts of Br were detected in most of the prostate samples and in nearly half of the liver samples. The data indicate that the Fe content decreases in malignant liver tissues, and the quantities of Cu and Zn are also decreased but to a lesser extent. Although these results are preliminary, they indicate that a large variation in trace elements exists among individuals with the same disease. Since this variation is not attributable to experimental error, it apparently reflects true individual differences.

Walter et al.[67] used 3-MeV protons to excite X-ray emission from tissue samples, such as body fluids, ion-exchange membranes, proteins, and others. A linear response was demonstrated for such elements as Pb, Cu, Zn, Co, and Mn from 5 ng to greater than 2 mg, with a lower limit of sensitivity of about 200 pg.

Van Rinsvelt et al.[68] have investigated trace elements in human disease by analyzing approximately 1500 samples of ashed human tissues (mostly from autopsies) by proton-induced X-ray emission analysis. Up to 15 different organs from each autopsy were used in the investigation as well as a variety of disease conditions, including neoplasis, chronic degenerative diseases, arteriosclerosis, and metabolic and inflammatory diseases. In each organ, an average of 12 trace elements (with atomic numbers equal to or larger than 19) was detected and quantitative measurements were made for several elements, including K, Ca, Ti, Mn, Fe, Cu, Zn, Pb, Se, Br, Rb, Sr, Cd, Sn, and Ba.

Trace-element composition of hair has recently received much attention. Hair, because of its growth, can reflect the biomedical and environmental history of the subject and is convenient to handle and sample. Because hair accumulates trace elements over a period of time and is relatively inactive metabolically, it may offer information about the nutrient regime of the body. Efforts have been made to relate the trace-element content of human hair to some diseases that are known to influence trace elements in other parts of the body. Proton-induced X-ray emission spectroscopy allows quick measurements to be made of a number of elements in hair samples. Valkovic[1] indicates that the typical X-ray spectrum resulting from bombarding hair with 3-MeV protons contains peaks associated with K, Ca, Ti, Mn, Fe, Ni, Cu, Zn, Pb, As, Se, and Br. Elemental concentration of these elements can be easily measured.

A great interest in the serum concentrations of numerous trace elements has developed in the biochemical sciences. Blood samples contain trace quantities of Cu, Fe, Al, Ba, Mn, Ni, Cs, Sn, Sr, Cr, Zn, Pb, Mo, Cd, and others, but only Cu, Fe, Sr, and Zn appeared in 100% of the specimens investigated.[1] These elements attracted considerable interest, since their concentrations seem to hold much promise as a clinical test in several pathological conditions. Charged-particle induced X-ray emission spectroscopy makes it possible to detect simultaneously several trace elements, and it has the advantage that very small quantities of serum are required in target preparation.

A technique for Se determination in blood serum using X-ray emission spectroscopy induced by 1.8 to 4.0-MeV protons has been described by Berti et al.[69] A sensitivity of less than 0.01 ppm for a 100-min counting time (1.8-MeV protons) and for a 30-min counting time (4.0-MeV protons) was obtained.

Bearse et al.[70] have described elemental analysis of whole blood using proton-induced X-ray emission. Samples of 0.1 ml whole blood from humans were dried, weighed, and then ashed. Drops made from the ash and a 400-ppm Pd solution were placed onto Formvar[R] backings. The samples were irradiated with 2.25-MeV protons, and Fe, Cu, Zn, Se, and Rb were detected with a precision of 7, 18, 7, 50, and 19%, respectively, in human whole blood.

The use of muons for elemental analysis may have applications in biological materials. The spectrum of X rays emitted when negative muons are captured in biological material provides quantitative information about the elemental composition and, to a lesser extent, qualitative information about the molecular composition of the material. Daniel[71] and Rosen[72] have pointed out the usefulness of this technique in clinical diagnosis and medical research. Muonic X-ray spectra from normal and pathological tissues reflect the existence of significant differences in elemental composition. The X-ray technique may also be very useful for monitoring elemental changes occurring during the course of disease management. The muonic spectral measurements of Hutson and associates[73] on dog's blood clearly indicated peaks for C, N, and O and could be used to determine amounts of them and other low atomic number elements present. The researchers used a thin Ti box so that muonic X rays originating from the box would not interfere with those of the sample material.

A potentially useful application of muonic X rays is in the elemental analysis of tissue deep within the body, thereby eliminating the need for surgical intervention. Hutson et al.[73] discussed a system that could be used for X-ray analysis of a volume of tissue within the body; a schematic diagram of this system is shown in Figure 13.

Valkovic[1] describes many other investigations from a large list of references.

4.3. Electron Excitation

4.3.1. DESCRIPTION

Fluorescence analysis using electron excitation has special character-istics that set it apart from fluorescence methods employing other forms of

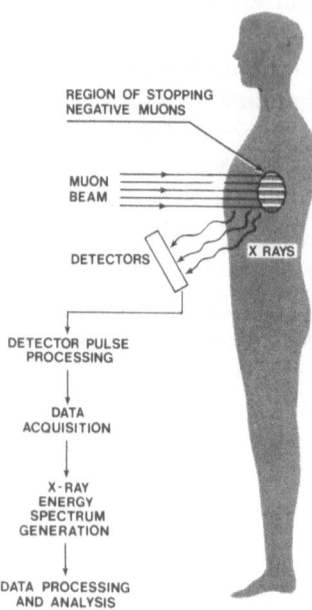

REGION OF STOPPING
NEGATIVE MUONS

MUON
BEAM

DETECTORS   X RAYS

DETECTOR PULSE
PROCESSING

DATA
ACQUISITION

X-RAY
ENERGY
SPECTRUM
GENERATION

DATA PROCESSING
AND ANALYSIS

Figure 13. Schematic representation of a muon X-ray analysis system for tissue volumes deep within the body (Hutson *et al.*, 1976).

excitation. The electron systems use accelerating columns of electron microscopes that have the unique capability of providing highly focused electron beams. These beams form an extremely well-defined probe that can be used to analyze very small portions of a sample. Thus, electron beam devices are described as microanalysis systems in contrast to those previously described, which are more aptly called bulk analysis systems.

Electron excitation of characteristic X rays involves the classical electrostatic interaction between the incident electron beam and the inner orbital electrons of the target atoms. The result is specific shell ionization and subsequent fluorescence X-ray emission. As a function of incident electron energy, the K-shell ionization cross sections rise from the threshold for the given atom to a maximum of 2.5 to 4 times the binding energy and gradually decrease beyond that point.[23, 74] Figure 14 shows the K-shell ionization cross section as a function of electron energy (or accelerating voltage) for several elements.[23] The maximum cross sections have about the same magnitudes as those for proton excitation (see Figure 10) but are about an order of magnitude less than the peak photoelectric cross sections for the corresponding elements (see Figure 2). As with heavier charged particles, the energy dependence of the cross sections is more gradual than for photons. In practice, the total output of excitation intensity will tend to increase with increasing voltage because the electron gun "brightness" increases with voltage. However, because of the rise in background radiation with voltage,

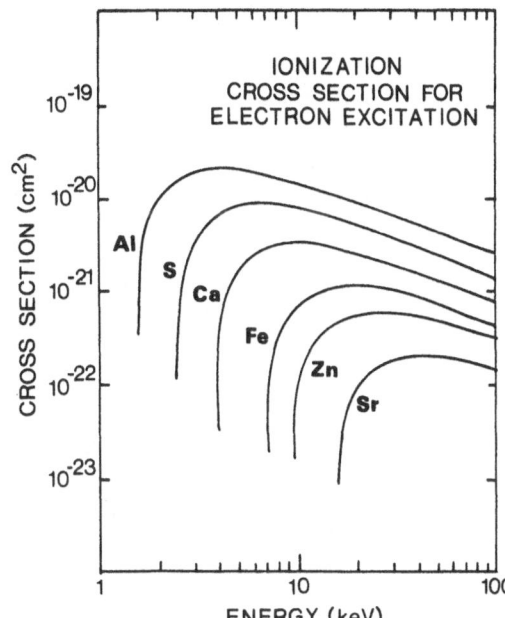

Figure 14. *K*-shell vacancy production cross sections as a function of electron energy (Jaklevic, 1979). Copyright 1979, the Chemical Rubber Company.

the best operating point for optimal detection limits will generally be below the maximum voltage available.[75]

Perhaps the major limitation in using electron excitation techniques for analytical trace-element investigation is the production of *background bremsstrahlung radiation*. This is due to the continuous deceleration process involving the interaction of electrons with the nuclear fields. The intensity of this bremsstrahlung emission (which represents white radiation not characteristic of the atoms under analysis) increases with the atomic number Z of the material and as the square of the accelerating voltage. The resulting spectrum exhibits the desired fluorescence (characteristic) peaks riding on a broad background, which provides a relatively poor signal-to-noise ratio when compared to photon or heavy charged-particle excitation. Figure 15 shows an example of this effect for electron excitation of the same sample in Figure 3.[23]

The particular advantage of electron excitation devices is their ability to analyze very small regions. This characteristic in part compensates for low signal-to-noise (sensitivity) properties and becomes important when studying trace elements that are generally distributed in a nonuniform manner. The spatial resolution will depend on the sample preparation, in the sense that for thick or bulk samples, the diffusion and scattering of electrons will result in a "teardrop" shaped excitation volume of dimensions considerably greater than the incident beam diameter. For ultra thin specimens, the excitation

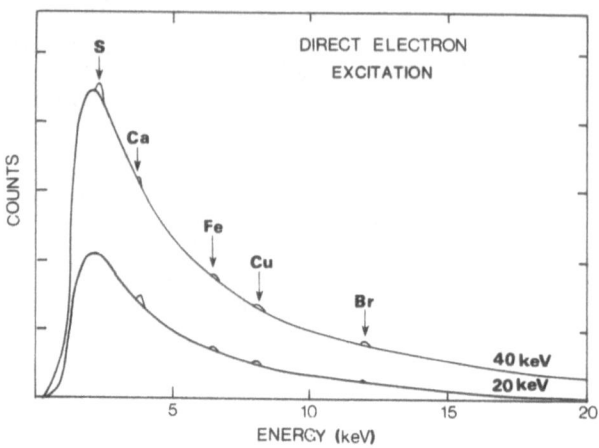

Figure 15. Idealized spectrum representing elements at 10 ppm in a hydrocarbon matrix excited by 20- and 40-keV electrons (Jaklevic, 1979). Copyright 1979, the Chemical Rubber Company.

volume is roughly cone shaped with minimal extension beyond the original beam boundaries as seen in Figure 16.[75]

The highly focused electron beam, coupled with the ability of some microscopes to scan, provides a method for obtaining qualitative *maps* of locations of a specific element in a sample. Quantitative analysis using electron beam systems requires a few special considerations. The elemental concentration will be proportional to the specific X-ray intensity divided by the appropriate correction factors. A discussion of three of the most important processes involved in the model of electron excitation X-ray fluorescence follows. These processes involve the so-called "ZAF" model and

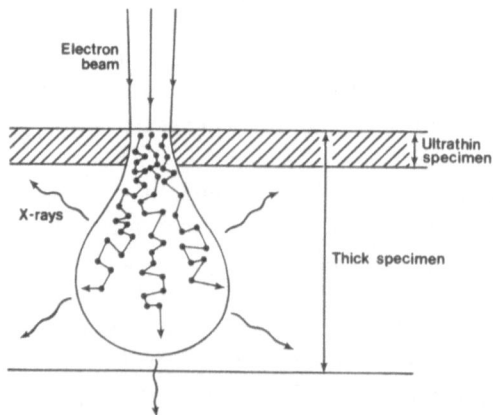

Figure 16. Diffusion of electrons entering the surface of a thick specimen (Chandler, 1978).

include effects of atomic number (Z), absorption (A), and secondary fluorescence (F).

The atomic-number effect is associated with penetrating and scattering of the electron within the excitation volume. The correction factor is composed of the ratio of two terms R/S where R is the fraction of incident electron energy that remains in the sample and generally decreases with both increasing atomic number and increasing electron-accelerating voltage. The S is associated with the rate of electron energy loss (electron-stopping power) and is greatest for low atomic number matrices and low electron energies. Errors involved in estimating the atomic number factor are actually increased when extrapolating from the assumptions made for heavier materials to biological tissues. The fact that R and S vary in a manner that tends to cancel individual term errors, provides some relief given the overall accuracy. Russ[74] indicates that it is desirable to keep this factor in the 0.5–1.8 range (preferably 0.8–1.25), which generally limits the accelerating voltage to values no lower than about two times the absorption edge energy and no higher than seven to ten times that energy.

The absorption correction refers to self-absorption of those fluorescence X rays generated within the sample that consequently do not reach the detector. The absorption coefficient will depend on the accelerating voltage used, the X-ray energy, and the specimen composition. Iterative techniques are used involving compositional assumptions that define the attenuation followed by the subsequent calculational results and comparisons with reduced X-ray output intensity. An absorption factor less than about 0.5 indicates that the accelerating voltage is too high, resulting in excessive penetration depths that lead to analysis error.

Finally, secondary fluorescence, which tends to increase a specific element's X-ray intensity, may be produced as a result of absorption of the fluorescence X rays of a higher atomic-number element in the matrix. This factor will be increased if there are elements present with absorption edges just below the energy of the principle emission lines of another element. In practice, the low concentration of the trace elements under analysis in thin biological samples make secondary fluorescence negligible.

In broad terms, electron excitation analysis involves incorporating an X-ray fluorescence detector into the traditional electron microscope's optical system. Initially, electron microscopes were categorized as scanning electron microscopes (SEM) or transmission electron microscopes (TEM). More recently, design changes have provided the SEM system with transmission capability (called STEM) and vise versa, so that the two categories have tended to merge. Russ[74] has reviewed some of the basic comparative characteristics of these systems, which include:

1. Image resolution for most commercial SEMs (or STEMs) is in the range 5–20 nm, whereas the TEM may achieve a resolution of under 0.5 nm.

2. For the SEM, the normal image is formed through point-by-point collection of the low-energy secondary electrons emitted from a thin layer near the sample surface. Close to the surface, the incident beam electrons have not experienced any significant transverse scattering, and the secondary electrons provide a spatial resolution roughly comparable to the beam diameter. This method of imaging coupled with electronic signal amplification gives STEM images enhanced contrast relative to TEM images. It is noted that for bulk samples, the X-ray photon image would be of much lower spatial resolution than the secondary electron image, since the X-ray production volume's diameter is considerably greater than the incident beam dimension.

3. The SEM offers advantages in specimen chamber design that include (a) more chamber space, facilitating positioning of the X-ray detector relative to the sample and thus improving efficiency; (b) room for specially staged mounts for either thin samples designed from low atomic number materials (beryllium or graphite) that will reduce background, or specimens to be held at low temperatures.

4. In general, the TEM design incorporates a more intense electron beam and provides a higher accelerating voltage. Higher voltages may contribute to analytical error as a result of increased background due to scattered electrons; however, these voltages may be required for penetrating thicker samples.

5. The TEM is more flexible in terms of the electron optical system and facilitates operator control and monitoring the electron beam.

A review of TEM microanalysis with ultrathin biological samples by Chandler[75] further describes the terminology and design of various systems, such as (1) EMMA (electron microscope microanalyzer); these instruments incorporate twin wavelength-dispersive crystal spectrometers along with energy-dispersive analyzers to provide high trace element detection sensitivity. (2) TEAM (transmission electron analytical microscope); these systems generally use solid-state energy-dispersive detectors. (3) TEPA (transmission electron probe analyses); these systems were similar to the EMMA but of an earlier vintage. The electron image resolution was of low quality, and its primary function was examining and analyzing bulk samples (as opposed to thin-specimen analysis). Figure 17 shows a schematic of a TEM system.

For most of the elements in the periodic table from atomic number 11 (Na) to higher values, the minimum detectable concentration for electron excitation X-ray fluorescence is in the range 0.1–0.01% for bulk materials provided the accelerating voltage is high enough for efficient excitation. As already discussed, the fundamental limitation is the background bremsstrahlung radiation. For lower atomic-number elements, the detection limit is somewhat higher because such elements as C, N, and O emit fewer

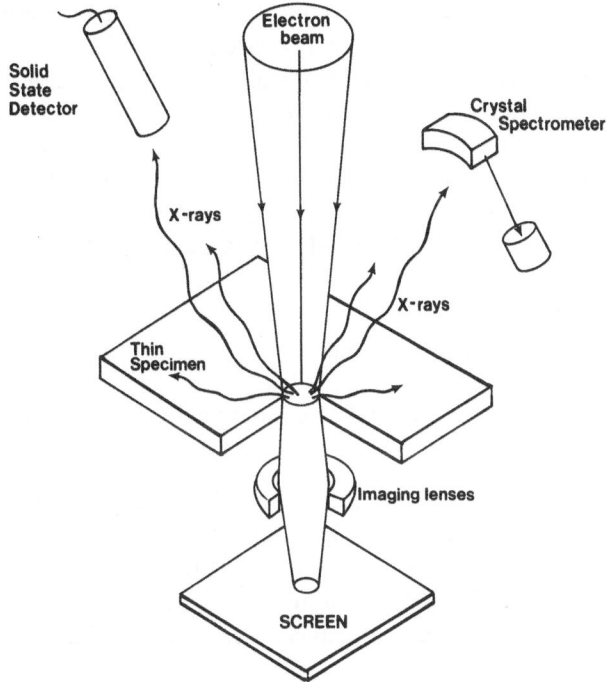

Figure 17. Schematic representation of a TEM system (Chandler, 1978).

characteristic X rays, and they are emitted at lower energies where they are more strongly absorbed.

For bulk biological tissue samples having a total excited volume in the range 10–50 μm^3, detecting trace elements within that volume becomes exceedingly difficult. Theoretically, thin-sample preparation allows detection limits to be brought down to the few-hundred-parts-per-million level (0.01%) for analyzed volumes in the range 10^{-3}–10^{-4} μm^3, which would represent an elemental mass as small as 10^{-20} g (a few hundred atoms) for a nominal density of 1 g/cm^3. In practice, minimum detectable quantities are 10^{-17} to 10^{-18} g with an analysis time of several minutes.[74] Detection limits could be increased by using higher incident electron beam intensity; however, the ultimate limitation would involve the ability of the sample to withstand heating effects at the higher current densities.

4.3.2. SAMPLE PREPARATION

Sample preparation techniques have been reviewed by Russ[74] and Chandler.[75] Sample preparation depends acutely on the intended purpose of analysis. The proper preparation of the biological tissue specimen is critical and represents one of the major technological aspects of successful X-ray analysis. In brief, techniques include (1) conventional fixation (aldehyde or alcohol) and embedding, (2) histochemical methods, (3) freeze substitution (replacing tissue water with resin), (4) freeze drying and embedding, (5) cryo-ultramicrotomy, (6) air drying, and (7) ashing and replication. Details and further references concerning these methods are provided by Chandler.[75] In addition, the important area of tissue-equivalent fluorescence standards required for quantitative analysis of thin specimens is discussed in this reference.

4.3.3. MEDICAL APPLICATIONS

The X-ray fluorescence analysis of biological samples may be classified into four broad categories[74, 75]:

1. Naturally occurring elements—localization and analysis (endogenous electrolytes, enzyme cofactors, metals, etc.).
2. Detecting elements associated with pathological conditions or accidently introduced matter (foreign particles, toxic substances).
3. Detecting deliberately introduced foreign material (tracers, drugs, stains, labels, etc.).
4. Histochemical reaction products (molecular group localization, etc.).

By far, most of the investigative analysis to date has involved detecting of trace endogenous elements in tissue. However, most of the studies have been qualitative in nature rather than quantitative. The following is an abbreviated list of representative work across a broad spectrum: Ahmed[76] has used EMMA instrumentation for investigating P and Ca in human breast carcinoma; Ashraf et al.[77, 78] have used STEM techniques to study P and Ca in myocardium muscle, while Yarom et al.[79] have looked at Cl, K, and Ca in the myocardium as a function of sample preparation. Davies et al.[80] have applied TEAM techniques in analyzing Na, Mg, P, S, K, and Ca in heart tissue; Chandler[81] has analyzed human chromosomes for Mg, P, S, Ca, Fe, and Zn using an EMMA. Kidney studies have been performed by Dempsey et al.[82] using a STEM and by Baker and Appleton[83] using a TEAM to analyze P, S, Cl, Ag, Mn, K, Ca, Fe, and Si. Cartilage and bone studies have been carried out by Ghadially et al.[84] using a STEM to analyze P, K, and Fe; and by Murphy et al.[85] using a TEAM to analyze Fe.

A wide range of additional tissues and organs have been subjected to electron excitation fluorescence analysis. they include, in part, ovary (EMMA—Mg, Si);[86] endometrium (TEAM—Cu);[87] sperm (TEAM—Zn);[88] stomach (EMMA—Mg, Al, Si, P, S, Cl, K, Ti, Fe);[89] pancreas (EMMA—Na, Mg, K, Ca, Ma, Sb);[90] cornea (EMMA—Na, K);[91] brain (TEAM—S, Co);[92] blood (EMMA—P, S, Cl, K, Ca);[93] and skin (TEAM—Si, P, S, Cl, K, Ca).[94]

Chandler[75] has made an extensive bibliographical review that includes medical investigations. His review table provides information about the instrumentation used, sample preparation techniques, type of tissue, and elements under investigation.

5. SUMMARY

It is interesting to compare the different modes of excitation in terms of precision, accuracy, and minimum detectable limits. Electron excitation is a specialized microanalytical procedure in which high accuracies are not usually achieved. For homogeneous samples, minimum detectable limits of less than 100 ppm are not practical.[23] The advantage of this technique is its ability to analyze extremely small areas of a sample and yield information regarding relative spatial distributions of elements in the sample. It is difficult to compare meaningfully heavy charged-particle excitation and photon excitation because of all the variables involved, such as the type of source, the sample form, the particular element of interest, etc. However, both methods are, in general, capable of analyzing to the 1-ppm level.

Because charged-particle beams can be focused, a smaller amount of the sample material is necessary with this type of analysis than with photon excitation. Charged-particle beams are also more effective for analyzing low atomic number elements ($12 \leqslant Z \leqslant 22$).[1] However, photon sources (X-ray tubes, radionuclides) are much cheaper and much more convenient than charged-particle accelerators, which may, indeed, be the overriding considerations in choosing a method of fluorescence analysis.

In general, fluorescence analysis is a relatively fast, easy to use, multi-element analysis technique that can be used to detect all elements of atomic number greater than 11 over a wide range of elemental concentrations, from high percentages to trace quantities in the fractional ppm range. In general, the technique is non destructive, and the sample may be used again for subsequent analyses (although charged-particle beams may be destructive to the sample). There is usually less sample preparation involved in fluorescence analysis than with other analytical techniques, and the sample may take one of many different forms. These features have resulted in the wide acceptance of X-ray fluorescence techniques in many medical applications.

REFERENCES

1. V. Valkovic, *Analysis of Biological Material for Trace Elements Using X-Ray Spectroscopy*, CRC, Boca Raton, Florida (1980).
2. E. J. Underwood, *Trace Elements in Human and Animal Nutrition*, Academic, New York (1971).

3. W. G. Hoekstra, J. W. Suttie, H. E. Ganther, and W. Mertz, ed., *Trace Element Metabolism in Animals V. 2*, University Park, Baltimore (1974).

4. I. L. Mulay, R. Roy, B. E. Knox, N. H. Suhr, and W. E. Delaney, Trace-metal analysis of cancerous and noncancerous human tissues, *J. Nat. Cancer Inst.* **47**, 1–13 (1971).

5. M. H. Seltzer, F. E. Rosato, and M. J. Fletcher, Serum and tissue magnesium levels in human breast carcinoma, *J. Surg. Res.* **10**, 159–162 (1970).

6. A. Danielsen and E. Steinnes, A study of some selected trace elements in normal and cancerous tissue by neutron activation analysis, *J. Nucl. Med.* **11**, 260–264 (1970).

7. J. M. Janes, J. T. McCall, and L. R. Elveback, Trace metals in human osteogenic sarcoma, *Mayo Clin. Proc.* **47**, 476–478 (1972).

8. J. Sherman, The theoretical derivation of fluorescent X-ray intensities from mixtures, *Spectrochim. Acta* **7**, 283–306 (1955).

9. J. Sherman, Simplification of a formula in the correlation of fluorescent X-ray intensities from mixtures, *Spectrochim. Acta* **15**, 466–470 (1959).

10. H. J. Lucas-Tooth and B. J. Price, A mathematical method for the investigation of interelement effects in X-ray fluorescent analysis, *Metallurgica* **64**, 149–152 (1961).

11. H. J. Lucas-Tooth and C. Pyne, The accurate determination of major constituents by X-ray fluorescent analysis in the presence of large interelement effects, in *Advances in X-Ray Analysis* vol. 7 (W. M. Mueller, G. Mallet, and M. Fay, eds.), pp. 523–541, Plenum, New York (1964).

12. J. W. Criss and L. S. Birks, Calculation methods for fluorescent X-ray spectrometry, *Anal. Chem.* **40**, 1080–1086 (1968).

13. A. Stephenson, Theoretical analysis of quantitative X-ray emission data: glasses, rocks, and metals, *Anal. Chem.* **43**, 1761–1764 (1971).

14. C. J. Sparks, Jr., Quantitative X-ray fluorescent analysis using fundamental parameters, in *Advances in X-Ray Analysis* vol. 19 (R. W. Gould, C. S. Barrett, J. B. Newkirk, and C. O. Ruud, eds.), pp. 19–52, Kendall/Hunt, Dubuque, Iowa (1976).

15. R. P. Gardner and A. R. Hawthorne, Monte Carlo simulation of the X-ray fluorescence excited by discrete energy photons in homogeneous samples including tertiary interelement effects, *X-Ray Spectrom.* **4**, 138–148 (1975).

16. A. R. Hawthorne and R. P. Gardner, Monte Carlo simulation of X-ray fluorescence from homogeneous multielement samples excited by continuous and discrete energy photons from X-ray tubes, *Anal. Chem.* **47**, 2220–2225 (1975).

17. R. P. Gardner, L. Wielopolski, and J. M. Doster, Adaption of the fundamental parameters of Monte Carlo simulation to EDXRF analysis with secondary fluorescer X-ray machines, in *Advances in X-Ray Analysis* vol. 21 (C. S. Barrett, D. E. Leyden, J. B. Newkirk, and C. O. Ruud, eds.), pp. 129–142, Plenum, New York (1978).

18. R. Woldseth, *X-Ray Energy Spectroscopy*, Kevex Corp., Burlingame, California (1973).

19. D. C. Camp, Physical principles, in *Medical Applications of Fluorescent Excitation Analysis* (L. Kaufman and D. C. Price, eds.), pp. 3–27, CRC Press, Boca Raton, Florida (1979).

20. J. M. Jaklevic and F. S. Goulding, Instrumentation for energy dispersive X-ray fluorescence, in *Medical Applications of Fluorescent Excitation Analysis* (L. Kaufman and D. C. Price, eds.), pp. 29–47, CRC, Boca Raton, Florida (1979).

21. Technical Data—Silicon (Li) X-Ray Detectors, E. G. and G. Ortec, Oak Ridge, Tennessee (December, 1977).

22. P. B. Hoffer, R. N. Beck, and A. Gottschalk, eds., *The Role of Semiconductor Detectors in the Future of Nuclear Medicine*, Society of Nuclear Medicine, New York (1971).

23. J. M. Jaklevic, Excitation methods for energy-dispersive analysis, in *Medical Applications of Fluorescent Excitation Analysis* (L. Kaufman and D. C. Price, eds.), pp. 49–67, CRC, Boca Raton, Florida (1979).

24. L. T. Dillman and F. C. von der Lage, *Radionuclide Decay Schemes and Nuclear Parameters*

for Use in Radiation-Dose Estimation, NM/MIRD Pamphlet no. 10, Society of Nuclear Medicine, New York (1975).

25. Radiation Sources, New England Nuclear Catalog 7M678A-1270, Boston, p. 5 (June, 1978).

26. T. J. Kneip and G. R. Laurer, Isotope excited X-ray fluorescence, *Anal. Chem.* **44**, 57A–68A (1972).

27. E. Vañó and L. González, Importance of geometry in biological sample analysis by X-ray fluorescence, *Med. Phys.* **5**, 400–403 (1978).

28. P. A. Feller, Determination of the suitability of photon-induced X-ray fluorescence analysis for the quantitation of selected low-atomic-number trace elements in biological materials, Ph.D. Dissertation, University of Cincinnati, Cincinnati, Ohio (1980).

29. D. A. Gedcke, E. Elad, and P. B. Denee, An intercomparison of trace element excitation methods for energy-dispersive fluorescence analyzers, *X-Ray Spectrom.* **6**, 21–29 (1977).

30. P. S. Ong, P. K. Lund, C. E. Litton, and B. A. Mitchell, An energy dispersive system for the analysis of trace elements in human blood serum, in *Advances in X-Ray Analysis* vol. 16 (L. S. Birks, C. S. Barrett, J. B. Newkirk, and C. O. Ruud, eds.), pp. 124–133, Plenum, New York (1973).

31. P. S. Ong, Trace elements in medicine, in *Medical Applications of Fluorescent Excitation Analysis* (L. Kaufman and D. C. Price, eds.), pp. 71–88, CRC, Boca Raton, Florida (1979).

32. P. S. Ong and H. L. Cox, Jr., Line-focusing X-ray monochromator for the analysis of trace elements in biological specimens, *Med. Phys.* **3**, 74–79 (1976).

33. T. G. Dzubay, B. V. Jarrett, and J. M. Jaklevic, Background reduction in X-ray fluorescence spectra using polarization, *Nucl. Instr. Meth.* **115**, 297–299 (1974).

34. L. Kaufman and D. C. Camp, Polarized radiation for X-ray fluorescence analysis, in *Advances in X-Ray Analysis* vol. 18 (W. L. Pickles, C. S. Barrett, J. B. Newkirk, and C. O. Ruud, eds.), pp. 247–258, Plenum, New York (1975).

35. R. H. Howell, W. L. Pickles, and J. L. Cate, X-ray fluorescence experiments with polarized X rays, in *Advances in X-Ray Analysis* vol. 18 (W. L. Pickles, C. S. Barrett, J. B. Newkirk, and C. O. Ruud, eds.), pp. 265–277, Plenum, New York (1975).

36. L. Kaufman, D. Shosa, and D. C. Camp, A high intensity source of polarized X rays for fluorescent excitation analysis, *IEEE Trans. Nucl. Sci.* **NS-24**, 525–531 (1977).

37. R. W. Ryon, Polarized radiation produced by scatter for energy dispersive X-ray fluorescence trace analysis, in *Advances in X-Ray Analysis* vol. 20 (H. F. McMurdie, C. S. Barrett, J. B. Newkirk, and C. O. Ruud, eds.), pp. 575–590, Plenum, New York (1977).

38. P. Standzenieks and E. Selin, Background reduction of X-ray fluorescence spectra in a secondary target energy dispersive spectrometer, *Nucl. Instr. Meth.* **165**, 63–65 (1979).

39. J. F. Tinney, *In vivo* X-ray fluorescence analysis—concepts and equipment, in *Semiconductor Detectors in the Future of Nuclear Medicine* (P. B. Hoffer, R. N. Beck, and A. Gottschalk, eds.), pp. 214–229, Society of Nuclear Medicine, New York (1971).

40. L. Patomäki and H. Olkkonen, Determination of mineral density and structural inhomogeneity of trabecular bone *in vitro* by X-ray fluorescence line scanning, *Int. J. Appl. Rad. Isot.* **25**, 401–406 (1974).

41. R. Cesareo and D. Del Principe, Analysis of iron in blood using radioisotopic-excited X-ray fluorescence, *Med. Phys.* **1**, 163–164 (1974).

42. L. Ahlgren and S. Mattsson, An X-ray fluorescence technique for *in vivo* determination of lead concentration in a bone matrix, *Phys. Med. Biol.* **24**, 136–145 (1979).

43. L. Ahlgren and S. Mattsson, Cadmium in man measured *in vivo* by X-ray fluorescence analysis, *Phys. Med. Biol.* **26**, 19–26 (1981).

44. L. Kaufman and C. J. Wilson, Determination of extracellular fluid volume by fluorescence excitation analysis of bromine, *J. Nucl. Med.* **14**, 812–815 (1973).

45. D. C. Price, L. Kaufman, and R. N. Pierson, Jr., Determination of the bromide space in man by fluorescent excitation analysis of oral bromine, *J. Nucl. Med.* **16**, 814–818 (1975).

46. L. Kaufman, F. Deconinck, D. C. Price, P. Guesry, C. J. Wilson, B. Hruska, S. J. Swann, D. C. Camp, A. L. Voegele, R. D. Friesen, and J. A. Nelson, An automated fluorescent excitation analysis system for medical applications, *Invest. Radiol.* 11, 210–215 (1976).

47. D. C. Price, S. J. Swann, S. T. C. Hung, L. Kaufman, J. P. Huberty, and S. B. Shohet, The measurement of circulating red cell volume using nonradioactive cesium and fluorescent excitation analysis, *J. Lab. Clin. Med.* 87, 535–543 (1976).

48. A. A. Moss, L. Kaufman, and J. A. Nelson, Fluorescent excitation analysis: a simplified method of iodine determination *in vitro, Invest. Radiol.* 7, 335–338 (1972).

49. J. A. Nelson, Studies of the kinetics of X-ray contrast agents using fluorescent excitation analysis, in *Medical Applications of Fluorescent Excitation Analysis* (L. Kaufman and D. C. Price, eds.), pp. 129–135, Plenum, New York (1979).

50. L. Kaufman, D. Shames, and M. Powell, An absorption correction technique for *in vivo* iodine quantitation by fluorescent excitation, *Invest. Radiol.* 8, 167–169 (1973).

51. N. Alazraki, J. W. Verba, J. E. Henry, R. Becker, A. Taylor, and S. E. Halpern, Noninvasive determination of glomerular filtration rate using X-ray fluorescence, *Radiology* 122, 183–186 (1977).

52. P. Guesry, L. Kaufman, S. Orloff, J. A. Nelson, S. Swann, and M. Holliday, Measurement of glomerular filtration rate by fluorescent excitation of nonradioactive meglumine iothalamate, *Clin. Nephrol.* 3, 134–138 (1975).

53. R. Cesareo, D. Del Principe, G. Mancuso, and D. B. Tallarida, *In vitro* labelling of platelets with stable selenocystine, *Int. J. Appl. Radia. Isot.* 27, 324–326 (1976).

54. R. Cesareo, G. Tallarida, and F. Baldoni, Determination of hemodynamic parameters in the rabbit by X-ray fluorescence excitation, *Int. J. Appl. Radia. Isot.* 26, 285–289 (1975).

55. P. B. Hoffer, R. E. Polcyn, R. Moody, H. J. Lowe, and A. Gottschalk, Fluorescence detection: application to the study of cerebral blood flow, *J. Nucl. Med.* 10, 651–653 (1969).

56. R. A. Moody, P. B. Hoffer, R. E. Polcyn, H. J. Lowe, A. Gottschalk, and G. D. Dobben, K-shell fluorescence for the study of regional cerebral blood flow, *J. Neurosurg.* 35, 181–184 (1971).

57. M. F. Lubozynski, R. J. Baglan, G. R. Dyer, and A. B. Brill, Sensitivity of X-ray fluorescence for trace element determinations in biological tissues, *Int. J. Appl. Rad. Isot.* 23, 487–491 (1972).

58. H. L. Cox and P. S. Ong, Sample mass determination using Compton- and total scattered excitation radiaton for energy-dispersive X-ray fluorescent analysis of trace elements in soft tissue specimens, *Med. Phys.* 4, 99–108 (1977).

59. J. M. Jaklevic, W. R. French, T. W. Clarkson, and M. R. Greenwood, X-ray fluorescence analysis applied to small samples, in *Advances in X-Ray Analysis* vol. 21 (C. S. Barrett, D. E. Leyden, J. B. Newkirk, and C. O. Ruud, eds.), pp. 171–185, Plenum, New York (1978).

60. L. Kaufman, D. M. Shames, R. H. Greenspan, M. R. Powell, and V. Perez-Mendez, Cardiac output determination by fluorescence excitation in the dog, *Invest. Radiol.* 7, 365–368 (1972).

61. M. E. Phelps, R. L. Grubb, Jr., and M. M. Ter-Pogossian, *In vivo* regional cerebral blood volume by X-ray fluorescence: Validation of method, *J. Appl. Physiol.* 35, 741–747 (1973).

62. F. Folkman, Progress in the description of ion induced X-ray production: Theory and implication for analysis, in *Iron Beam Surface Layer Analysis* vol. 2 (O. Meyer, G. Linker, and F. Kappeler, eds.), Plenum, New York (1976).

63. L. R. Anspaugh, W. L. Robinson, W. H. Martin, and O. A. Lowe, *Compilation of Published Information on Elemental Concentrations in Human Organs in Both Normal and Diseased States*, UCLR-51013, Lawrence Radiation Laboratory Report, Berkeley, California (1976).

64. M. Dabek, N. A. Dyson, and A. E. Simpson, Quantitative applications of proton-induced X-ray emission analysis in the fields of medicine and biology, *Proceedings of the Annual Conference of the Microbeam Analysis Society*, Microbeam Analysis Society, Bethlehem, Pennsylvania (1977).

65. H. Kubo, Reproducibility of proton-induced elemental analysis in biological tissue sections, *Nucl. Instr. Meth.* **121**, 541–545 (1974).

66. H. Kubo, S. Hashimoto, A. Ishibashi, R. Chiba, and H. Yokota, Simultaneous determinations of Fe, Cu, Zn, and Br concentrations in human tissue sections, *Med. Phys.* **3**, 204–209 (1976).

67. R. L. Walter, R. D. Willis, W. F. Gutknecht, and J. M. Joyce, Analysis of biological, clinical, and environmental samples using proton-induced X-ray emission, *Anal. Chem.* **46**, 843–855 (1974).

68. H. A. Van Rinsvelt, R. D. Lear, and W. R. Adams, Human diseases and trace elements: investigation by proton-induced X-ray emission, *Nucl. Instr. Meth.* **142**, 171 (1977).

69. M. Berti, G. Buso, P. Colautti, G. Moschini, B. M. Stievano, and C. Tregnaghi, Determination of selenium in blood serum by proton-induced X-ray emission, *Anal. Chem.* **49**, 1313–1315 (1977).

70. R. C. Bearse, D. A. Close, J. J. Malanify, and C. J. Umbarger, Elemental analysis of whole body using proton-induced X-ray emission, *Anal. Chem.* **46**, 499–503 (1974).

71. H. Daniel, The muon as a tool for scanning the interior of the human body, *Nuclearmedizin* **8**, 311–319 (1969).

72. L. Rosen, Relevance of particle accelerators to national goals, *Science* **173**, 490–497 (1971).

73. R. L. Hutson, J. J. Reidy, K. Sprunger, H. Daniel, and H. B. Knowles, Tissue chemical analysis with muonic X-rays, *Radiology* **120**, 193–198 (1976).

74. J. C. Russ, Electron probe X-ray microanalysis—principles, in *Electron Probe Microanalysis in Biology* (D. A. Erasmus, ed.), pp. 5–36, Chapman and Hall, London (1978).

75. J. A. Chandler, The application of X-ray microanalysis in TEM to the study of ultrathin biological specimens—a review, in *Electron Probe Microanalysis in Biology* (D. A. Erasmus, ed.), pp. 37–93, Chapman and Hall, London (1978).

76. A. Ahmed, Calcification of human breast carcinomas: ultrastructural observations, *J. Path.* **117**, 247–251 (1975).

77. M. Ashraf and C. M. Bloor, X-ray microanalysis of mitochondrial deposits in ischaemic myocardium, *Virchows Archiv. B., Cell Path.* **22**, 287–298 (1976).

78. M. Ashraf, H. D. Sybers, and C. M. Bloor, X-ray microanalysis of ischaemic myocardium, *Exp. and Molec. Path.* **24**, 435–440 (1976).

79. R. Yarom, P. D. Peters, M. Scripps, and S. Rogel, Effects of specimen preparation on intracellular myocardial calcium, *Histochem.* **38**, 143–153 (1974).

80. T. W. Davies and A. J. Morgan, The application of X-ray analysis in the transmission electron analytical microscope (TEAM) to the quantitative bulk analysis of biological microsamples, *J. Microscopy* **107**, 47–54 (1976).

81. J. A. Chandler, X-ray microanalysis of human chromosomes, *Lancet* **7859**, 687 (1974).

82. E. W. Dempsey, F. J. Agate, M. Lee, and M. L. Purkerson, Analysis of submicroscopic structures by their emitted X rays, *J. Histochem. Cytochem.* **21**, 580–586 (1973).

83. J. R. Baker and T. C. Appleton, A technique for electron microscope autoradiography and X-ray microanalysis of diffusible substances using freeze-dried fresh-frozen sections, *J. Microscopy* **108**, 307–315 (1976).

84. F. N. Ghadially, A. F. Oryschak, R. L. Ailsby, and P. N. Mehta, Electron probe X-ray microanalysis of siderosomes in haemarthrotic articular cartilage, *Virchows Archiv. B., Cell Path.* **16**, 43–49 (1974).

85. M. J. Murphy and J. C. Piscopo, Cellular iron in aplastic anaemic human bone marrow: a study by energy-dispersive analysis of X rays, *J. Submicroscopic Cytol.* **8**, 269–276 (1976).

86. K. Griffiths, W. J. Henderson, J. A. Chandler, and C. A. F. Joslin, Ovarian cancer: some new analytical approaches, *Postgrad. Med. J.* **49**, 69–72 (1973).

87. A. Gonzalez-Angulo and R. Azner-Ramos, Ultrastructural studies on the endometrium of

women wearing T-Cu 200 IUDs by means of transmission and scanning EM and X-ray dispersive analysis, *Amer. J. Obs. Gyn.* **125**, 170–178 (1976).

88. J. A. Grimaud, J. C. Czyba, and N. Guillot, Energy-dispersive X-ray spectrometry of human spermatozoa in electron microscopy, *Comptes Rendus des Séances de la Société de Biologie* **170**, 1233–1236 (1977).

89. W. J. Henderson, D. M. D. Evans, J. D. Davies, and K. Griffiths, Analysis of particles in stomach tumours from Japanese males, *Environ. Res.* **9**, 240–249 (1975).

90. L. Herman, T. Sato, and C. N. Hales, The electron microscopic localization of cations to pancreatic islets of Langerhans and their possible role in insulin secretion, *J. Ultrastructure Res.* **42**, 298–311 (1973).

91. S. Hodson and J. Marshall, Tissue sodium and potassium: direct detection in the electron microscope, *Experientia* **26**, 1283–1284 (1970).

92. J. B. Kirkham, L. J. Goodman, and R. L. Chappel, Identification of cobalt in processes of stained neurons using energy spectra in the electron microscope, *Brain Res.* **85**, 33–37 (1975).

93. R. J. Skaer and P. D. Peters, The state of chlorine and potassium in human platelets and red cells, *Nature* **257**, 719–720 (1975).

94. K. Takaya, Intranuclear silicon detection in a subcutaneous connective tissue cell by energy-dispersive X-ray microanalysis using fresh air-dried spread, *J. Histochem. Cytochem.* **23**, 681–685 (1975).

4

Basic Imaging Properties of Radiographic Systems and Their Measurement

KUNIO DOI

1. INTRODUCTION

The physical image quality of radiographs is known to be affected by at least three fundamental factors, namely, contrast, resolution (or sharpness), and noise (or radiographic mottle).[1, 2] The contrast of a radiograph, commonly referred to as radiographic contrast, is related to the film contrast (or gradient), radiation contrast, and primary (or scatter) fraction. The gradient is usually derived from the characteristic curve (or H and D curve) of the screen-film system on which the radiographic image is recorded. The H and D curve is a basic imaging characteristic in radiographic systems and has been measured by a technique called sensitometry.

The resolution properties of a radiographic system or of its imaging components have been evaluated by means of the modulation transfer function (MTF), which describes the frequency response of an imaging component or of a total imaging system. Measurements of the MTFs of X-ray intensifying screens were first attempted about two decades ago;[3-6] this was probably the beginning of modern X-ray imaging science. The concept of frequency analysis was then applied to evaluating the noise of radiographic

KUNIO DOI • Kurt Rossmann Laboratories for Radiologic Image Research, Department of Radiology, University of Chicago, Chicago, Illinois 60637.

images by means of the Wiener spectrum. This type of analysis has significantly increased our understanding of radiographic mottle.[7-9]

In this chapter, methods and procedures for measuring the basic imaging characteristics of radiographic systems are described, and physical parameters that are pertinent to these measurements are discussed.

2. X-RAY SENSITOMETRY

When a radiographic image is produced on a photographic film, the distribution of the image appears as variations in the darkness of the film. The film darkening is often quantified by the photographic density or optical density, which is equal to the logarithm of the reciprocal of the transmittance of the film. Since the term *optical density* is used in solid-state physics and chemistry as an indicator of light absorption in materials, the term *photographic density* is preferred for the darkening of film in photography and radiography. For simplicity, we shall refer to it here as density.

The relationship between density and radiation intensity has usually been illustrated by plotting the logarithmic relative intensity on the abscissa; this is called a characteristic curve or H and D curve. The H and D curve has been used for a number of different studies, which include not only determining the film gradient but also measuring MTFs and many other physical properties of radiation intensities as well as monitoring films and film processing for quality assurance programs in diagnostic radiology departments.

It is known that the shape of H and D curves depends on such factors as the type of film, processing conditions, and exposure time. Specifically, the inapplicability of the reciprocity law to light exposure is an important factor for accurately measuring H and D curves, since the H and D curve of a screen-film system varies with the exposure time employed, whereas the H and D curve of a film under direct X-ray exposure without screens does not. Since the radiographic image is obtained by an X-ray exposure at a fixed exposure time, and H and D curve used for quantitative analysis must be determined by a method that employs an exposure time comparable to the actual time used for radiographic imaging. Therefore, only the radiation intensity is changed in deriving the H and D curve of a screen-film system. This procedure is called intensity-scale sensitometry, whereas time-scale sensitometry is employed when the reciprocity law holds.

2.1. Methods of Measurement

The simplest method of varying X-ray intensity is changing the distance between the X-ray tube focal spot and the screen-film system according to the inverse-square law. The first intensity-scale X-ray sensitometer was built by Cleare at the Eastman Kodak Research Laboratory;[10] later, a more sophisticated, automated X-ray sensitometer was constructed at the University of Chicago by Haus and Rossmann.[11] This sensitometer is still frequently used at the Rossmann Laboratory and is described in detail in Section 2.2. The inverse-square X-ray sensitometer is considered the most accurate device for measuring H and D curves of screen-film systems and is presently regarded as the standard.

Another method of varying X-ray intensity is step-wedge sensitometry, in which a step wedge made of absorbers, such as aluminum and/or copper, is used to attenuate the X-ray beam incident on the screen-film system.[12,13] To minimize the change in beam quality, a heavy filter, such as approximately 30 mm of aluminum is placed in front of the X-ray tube collimator. In addition, lead masks may be used to reduce the scatter from the step wedge.[14] The H and D curves measured by this method indicate a lower contrast and a greater latitude than corresponding values measured with an inverse-square sensitometer.[15]

The X-ray intensity may be varied by adjustments in the X-ray tube current. However, if a wide range of tube current is used, the kV wave form may change, thus resulting in inaccurate control of the X-ray intensity. According to our experience, however, this procedure offers a simple and reasonable approach to determining H and D curves over a relatively narrow range of intensities. With this method, the contrast tends to be slightly lower when sensitometry is performed over a wide range of intensities.

Recently, a number of investigators[14-17] attempted to construct the H and D curve from measurements over many separate segments of the curve, which is called "bootstrap sensitometry"; Yester et al.[16] employed a kV-time bootstrap technique. At a low initial kV setting, they made three exposures of a screen-film system by using exposure times of 0.25, 0.35, and 0.5 sec. By relating the relative exposure to the corresponding density, this provides a small lower portion of the H and D curve. They assumed that the reciprocity law holds for this experiment because the range of exposure times employed is small. The second set of exposures, with the same exposure times used for the first set, is then made at an increased kV setting in such a way that the lowest density in the second set is below the highest density in the first set; thus, the two segments of the H and D curve overlap. Additional exposures at increased kVs are then made, which complete the curve.

Bootstrap sensitometry can be performed by means of a combination of other parameters, such as mA and time instead of kV and time. Wagner et al.[15] employed an aluminum step wedge containing 25 steps by exposing twice at two different exposure times, which corresponds to a step-wedge–time combination. The use of the step wedge simplifies the experimental procedure considerably. Bednarek and Rudin[14] employed a step wedge–

distance combination and also a graphic determination of the H and D curve based on the relationship of densities between two films exposed at different distances. This graphical approach was intended to eliminate the uncertainties of constructing the H and D curve from segmental measurements. The authors found a close agreement between bootstrap and inverse-square sensitometry.

Another method of sensitometry that is widely used for quality assurance in processing conditions in radiology employs a simulated-light sensitometer. With this device, light intensity is adjusted by means of a neutral density filter through which a film is exposed. This method is fundamentally different from approaches based on adjusting the X-ray intensity. However, it is generally believed that the simulated-light sensitometer can produce the same H and D curve as that obtained with the X-ray sensitometer if the spectral distribution of the simulated light accurately matches the spectral distribution of light emitted from the intensifying screens and if the light-absorbing step wedge is accurately made. In our experience, H and D curves obtained with a well-made light sensitometer are very similar to those measured with the inverse-square sensitometer, especially at densities below about 2.0.

2.2. Inverse-Square X-Ray Sensitometry

One can determine the H and D curve of a screen-film system by exposing the system at various distances from the X-ray tube and by plotting the exposures at these positions versus the corresponding film densities. This can be done manually in various ways; however, precision, reproducibility, and efficiency are improved by automated procedures.

An automatic intensity-scale X-ray sensitometer designed by Haus and Rossmann[11,18] (Figure 1) is composed of (1) an X-ray source continuously emitting radiation at a constant intensity at diagnostic kilovoltages; (2) a shutter permitting accurate and reproducible timing of exposure; (3) a device providing automatic, stepwise increases of the focus-film distance to produce known, constant increments in exposure; and (4) a special cassette and cassette-moving device. The X-ray source is a conventional 250-kV General Electric Maxitron therapy X-ray tube. Several modifications were made so that the output of the tube could be kept stable; they included using a line voltage stabilizer and precise meters for monitoring and controlling kilovoltage, the primary line voltage, and the tube current setting. At 80 kV, the fluctuation in kilovoltage is at most $\pm\frac{1}{2}$ kV at the constant tube current. With these instruments, we found that it was not necessary to continuously monitor the X-ray beam in the film plane.

The timing shutter and cassette carriage are shown in more detail in Figure 2. Two superimposed lead-lined sector wheels that can be rotated

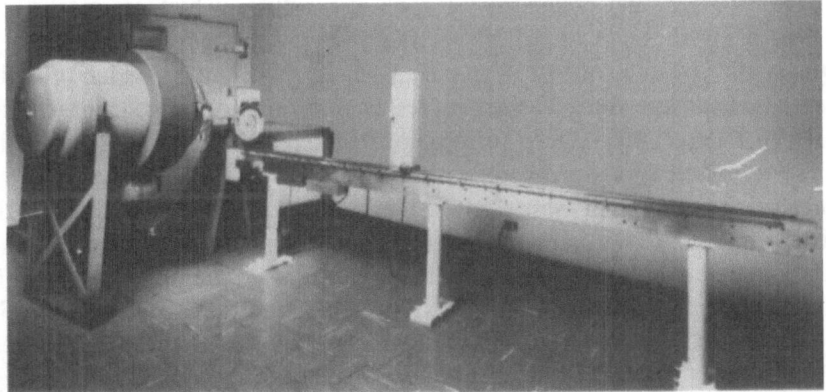

Figure 1. Intensity-scale X-ray sensitometer with a 250-kV therapy X-ray tube, timing-shutter wheel, and cassette carriage on the track assembly.

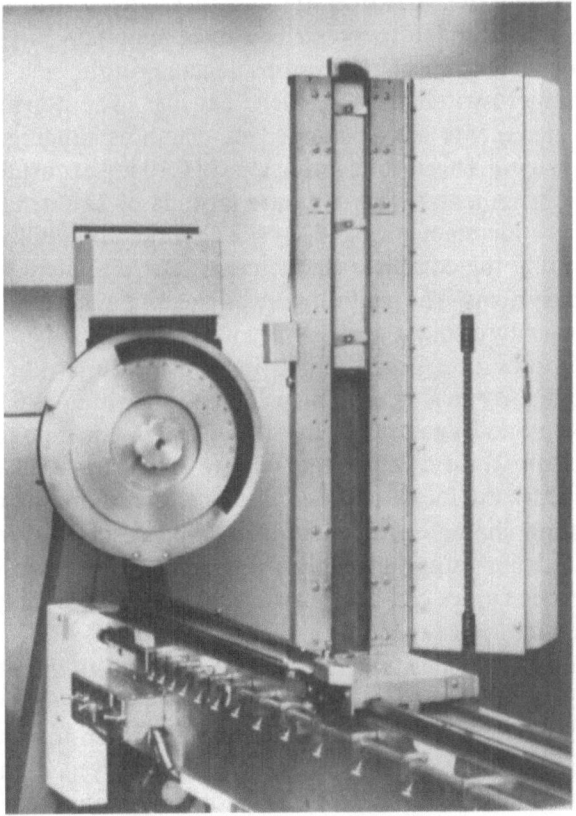

Figure 2. Timing-shutter wheel and cassette carriage.

relative to each other define the selection of apertures, which are marked off in degrees. With the sector wheels set at a given degree opening, the shutter is rotated at a constant rate of 18 rpm by a synchronous motor, which provides a specific exposure time. The exposures used in practice range from 0.05 sec (5.4°) to 1.48 sec (160°). Two springs, located behind the aperture defined by the sector wheel and controlled by solenoids, allow two lead shutters to open and close. The action of these shutters is synchronized with the rotation of the wheel and the movement of the cassette carriage by two microswitches positioned above the sector wheels. These shutters avoid film exposure while the cassette carriage is moving. A diaphragm behind the lead shutters limits the beam size. The size of this diaphragm was chosen so that (1) the film sees the entire focal spot at all focal spot–film distances; (2) the beam size is large enough at the shortest focal spot–film distance to provide a uniform exposure to the film; and (3) the beam size is limited so that no scatter is introduced by the beam striking the bench, wall, or objects other than the film, at even the greatest focal spot–film distance used.

A chain drive extending over the length of the bench automatically moves the cassette carriage to prescribed focal spot–film distances. These distances are defined by mechanical stops, located on the side of the bench, that activate a microswitch on the cassette carriage. The stops which define the exposure increments are positioned at distances ranging from 26 to 320 cm (10 to 126 in). This results in 23 steps of 0.10 log exposure increments (26% exposure increments). The exposure latitude of 23 steps consisting of 0.10 log exposure increments (log $E_{REL} = 2.2$) generally fulfills the requirements of providing the complete characteristic curve of medical X-ray film and screen-film systems. The calibration was checked against an X-ray time-scale sensitometer by using a direct X-ray film.

Figure 3 shows an exposed film strip with measured film densities in column 1; the density across each step is uniform within ± 0.01. The film density can be plotted against the logarithm of the corresponding relative exposure (column 2) or against the relative exposure itself (column 3). Column 4 indicates the focal spot–film distance corresponding to each step.

The resulting characteristic curve for a typical screen-type X-ray film exposed with calcium tungstate intensifying screens is shown in Figure 4. Such sensitometric curves can be reproduced within an accuracy of $\pm 2\%$ of the exposure (± 0.01 log exposure).

By using this X-ray sensitometer, we have investigated the effects of some parameters on the shape of H and D curves. First, we found[18,19] that the shapes of H and D curves of most blue- and green-sensitive radiographic films do not change significantly with the screens employed, even though it is well known that the H and D curve of a photographic film depends on the spectral composition of the light used.[20] Haus et al. examined the dependence of H and D curves of Kodak RP film and 3M XD film by using

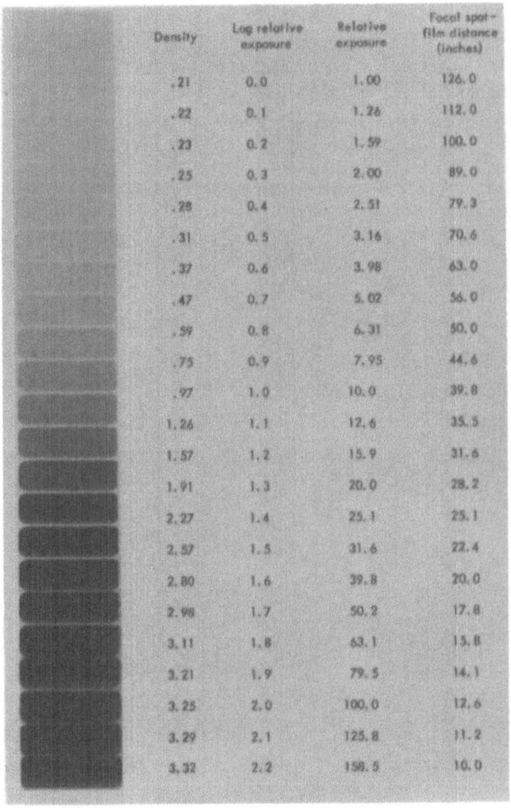

Density	Log relative exposure	Relative exposure	Focal spot - film distance (inches)
.21	0.0	1.00	126.0
.22	0.1	1.26	112.0
.23	0.2	1.59	100.0
.25	0.3	2.00	89.0
.28	0.4	2.51	79.3
.31	0.5	3.16	70.6
.37	0.6	3.98	63.0
.47	0.7	5.02	56.0
.59	0.8	6.31	50.0
.75	0.9	7.95	44.6
.97	1.0	10.0	39.8
1.26	1.1	12.6	35.5
1.57	1.2	15.9	31.6
1.91	1.3	20.0	28.2
2.27	1.4	25.1	25.1
2.57	1.5	31.6	22.4
2.80	1.6	39.8	20.0
2.98	1.7	50.2	17.8
3.11	1.8	63.1	15.8
3.21	1.9	79.5	14.1
3.25	2.0	100.0	12.6
3.29	2.1	125.8	11.2
3.32	2.2	158.5	10.0

Figure 3. Exposed film strip with corresponding measured film densities, log relative exposure, relative exposure, and focal-spot to film distance.

3M Alpha 8, Kodak X-Omatic Regular, and DuPont Par Speed screens, which are made of $(La, Gd)_2O_2S$, $(Ba, Sr)SO_4$, and $CaWO_4$ phosphors, respectively. The wavelengths of the light emitted by these screens differ considerably, ranging from green to ultraviolet. Nevertheless, there was no noticeable difference in the shapes of H and D curves when these screens were employed.

The dependence of X-ray beam quality on the shape of H and D curves was also studied by increasing the tube voltage to 120 kV and using a Thoreaus filter.[21,22] There was no significant change in the shapes of H and D curves from those obtained at 80 kV with a 0.5-mm Cu and a 3-mm Al filter, which was considered a representative beam quality in the medium diagnostic energy range. However, it is possible that, if a very soft X-ray beam is used, the shape of the H and D curve may change because of the difference in the fractions of light exposure in the front and back emulsions.

Major changes in the shapes of H and D curves can occur, however, depending on the film used and processing conditions,[23] including daily

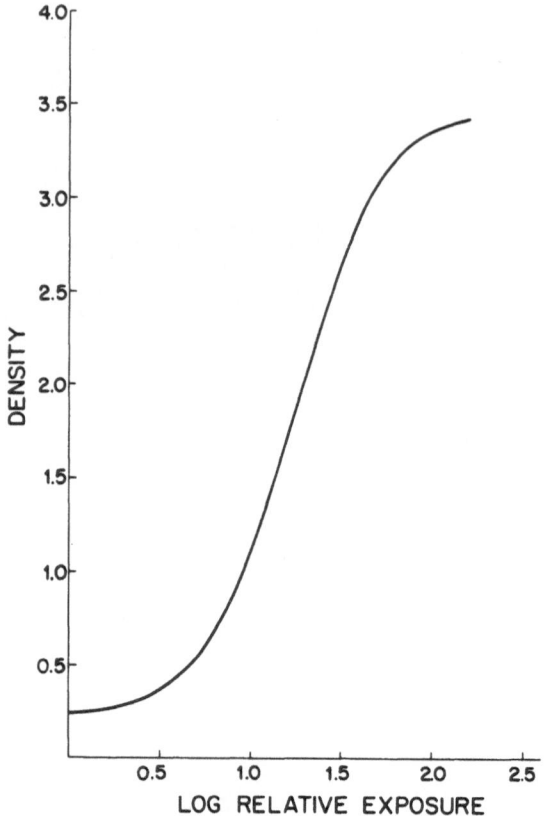

Figure 4. Characteristic (H and D) curve of a typical radiographic screen-film combination.

variations in the chemical condition of the developer and in the temperature in the film processor, and also batch-to-batch variations in the films as well as aging of films.

2.3. Application of a Curve-Smoothing Technique to the Determination of H and D Curves and Their Gradients

One of the important factors affecting radiographic image quality is the film contrast of the recording system, which may be expressed as the gradient of its H and D curve. The gradient G is defined by the slope of the H and D curve and is given by

$$G = \frac{\Delta D}{\Delta \log E} \tag{1}$$

where ΔD is a small density difference corresponding to an increment in the logarithm of the exposure. It should be noted that the gradient depends on the density and is sometimes called a point-by-point gamma.

From Eq. (1), the radiographic contrast ΔD, which is defined broadly as the difference in densities between two locations in a radiograph, can be described approximately by

$$\Delta D = 0.43 \, G \, \frac{\Delta E}{E} \tag{2}$$

where E and $E + \Delta E$ are the exposures responsible for the densities at the two locations. The quantity $\Delta E/E$ may be called the radiation contrast, which is related to the object being radiographed and the X-ray beam quality employed. It has been shown[21,155] that, for accurate estimates of radiographic contrast, the energy response of a recording system also has to be incorporated. When the scattered radiation is included in the final radiograph, as is usually the case, Eq. (2) may be extended to a more general expression that includes the scatter fraction S. The radiographic contrast ΔD_S is then described by

$$\Delta D_S = 0.43 \, G \, \frac{\Delta E}{E} \, (1 - S) \tag{3}$$

where the scatter fraction is equal to $E_S/(E + E_S)$ and E_S is the exposure due to the scatter. The scatter fraction depends on the energy response of the recording system.[156-158] Equation (3) shows that radiographic contrast is related in a simple way to the film gradient, radiation contrast, and scatter fraction. In most radiographic examinations, some of the scattered radiation coming from the object is usually removed by antiscatter techniques, such as the use of a grid, air gap, or scanning slit device.[159,160] When a grid is employed, the resulting scatter fraction is affected by many factors, such as geometric parameters of a grid as well as the specific imaging conditions used, which are discussed in greater detail in Refs. 161 and 162.

The gradient at each density is usually derived by means of numerical differentiating a H and D curve. Experimental determinations of H and D curves are subject to a relatively small error, which can, nevertheless, cause large fluctuations and uncertainties in the calculated gradient, since numerical differentiation tends to amplify errors. Large fluctuations in the gradient curve, which is a plot of the gradient versus the density, are common, and they introduce inaccuracy into the comparison of film contrasts.

In order to reduce this uncertainty, we have developed a curve-smoothing program for the PDP 8/e computer in our laboratory. With this program, H and D curve smoothing and the subsequent derivation of the gradient curve from the smoothed curve can be accomplished routinely. The program has proved useful and convenient as a part of the sensitometric evaluation of radiographic recording systems.[24]

One powerful technique that can be applied to numerical methods for fitting a smooth curve through measured data is estimating parameters by the principle of least squares. This method can reduce random errors to some extent and, in certain cases, it can even reduce systematic errors that may cause fluctuations in the data. Among the several methods of curve fitting by least squares, we have chosen orthogonal-polynomial smoothing for evenly spaced abscissas by using discrete Legendre polynomials[25] for H and D curves. This method has the following advantages: First, the discrete structure of polynomials in this form is suitable for computer do-loop manipulation without requiring numerical integration. Second, the orthogonality of the polynomials makes possible independent calculations of regression coefficients without having to solve for a system of simultaneous equations; thus, the problem of inverting a near-singular matrix, which occurs when the number of data points is large, is avoided.[26] Third, regression coefficients are independent of one another; to increase the order of regression by one or more, one can calculate the additional coefficient(s) independently without recomputing all of the previous coefficients, as is necessary in curve smoothing with ordinary polynomials. The form of the discrete Legendre polynomials used in the program and the derivation of the regression coefficients have been described previously.[24]

In this program, a polynomial of any desired degree can be used for smoothing. The ouput includes the fitted H and D curve, residuals, the residual root mean square (RRMS), gradients as a function of densities for both fitted and original curves, and the average gradient between the net densities 0.25 and 2.00. The fitted H and D curve and the gradient curve, as well as the original data points, are plotted on a CRT screen. It is then determined from the RRMS value and the CRT display whether the smoothing is satisfactory. The RRMS value is employed as a guide for fitting the curve. The final decision, however, is based on the visual comparison of the CRT display images of both H and D and gradient curves. This procedure is necessary because, in some cases, very high degree polynomials tend to follow fluctuating experimental data, resulting in obviously unsmooth curves (especially gradient curves) even with small RRMS values. If a polynomial of a different degree is desired, higher regression coefficients are calculated and new outputs given. This procedure is repeated until optimal curve smoothing is obtained.

The program was tested with H and D curves of many screen-film systems and direct X-ray films; some results are shown in Figures 5 and 6. The H and D curve of type AA X-ray film, fitted by fourth-degree polynomials, is shown in Figure 5(a). The gradient curve in Figure 5(b), derived from the fitted H and D curve, is smooth and fits the original gradient values well. In Figure 6(a), the curve for RP film with BG Hi-Speed screens was smoothed by sixth-degree polynomials. Some fluctuations in data points can be seen in

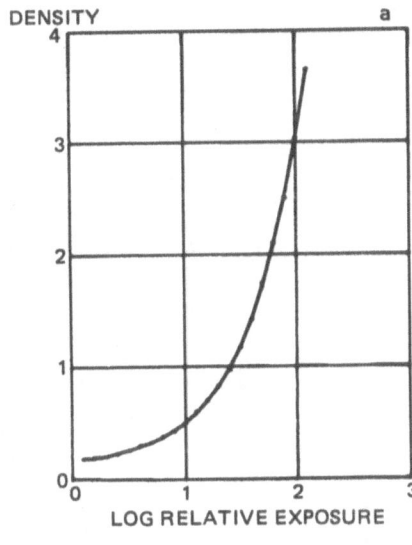

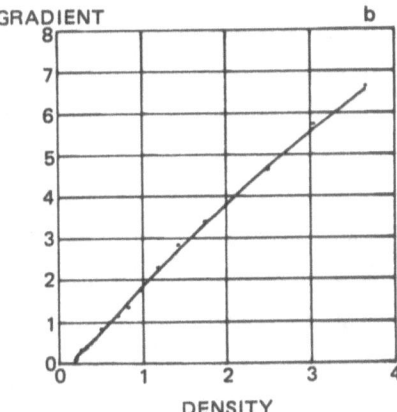

Figure 5. (a) H and D curve of Kodak Type AA X-ray film, smoothed by fourth-degree polynomials. (b) Gradient curve derived from the smoothed H and D curve of Type AA film. Points are gradient values obtained from an experimental H and D curve.

the middle density range. The gradient curve in Figure 6(b) reveals amplified fluctuations in the original data and the smoothly fitted curve.

We found that H and D curves of screen-film systems are, in general, smoothed optimally by sixth- or seventh-degree polynomials; some systems can be smoothed equally well with either degree. The eighth degree is too high for all H and D curves, since it tends to follow fluctuations in the data, especially at high densities. The H and D curves of direct X-ray films are fitted by polynomials of lower degree than needed for screen-film systems. Curve smoothing has been carried out routinely in our laboratory with a PDP 8/e digital computer after each H and D curve measurement. This procedure has led to improved sensitometry and evaluations of film characteristics.

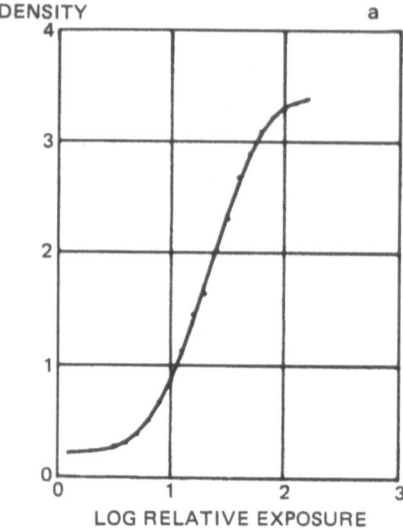

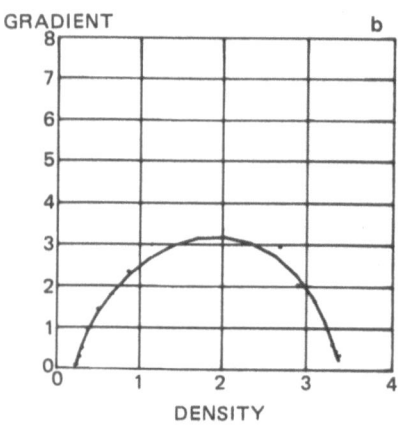

Figure 6. (a) H and D curve of Kodak XRP film with GAF BG Hi-Speed screens, smoothed by sixth-degree polynomials. (b) Gradient curve of XRP film with BG Hi-Speed screens.

3. RESOLUTION PROPERTIES OF RADIOGRAPHIC IMAGING SYSTEMS

The image degradation of an object in terms of its spatial distribution, which has been called "unsharpness" or "blur", is caused by the resolution properties of the imaging system used. The basic characteristic of resolution properties has been described by the system's response to a delta-function like input that is a point object in the two-dimensional case or a line object in the one-dimensional case. The spatial distributions of these responses are called the point spread function (PSF) and the line spread function (LSF); these functions are very similar to the impulse response that has been widely used in electronics. In fact, many descriptors of imaging performance in radiography, photography, and optics were derived from identical or similar concepts established in electronics and communication systems.

The Fourier transform of the PSF or the LSF has been called the optical transfer function (OTF), which describes the response of an imaging system in the spatial-frequency domain.

The OTF of an imaging system $H_2(u, v)$ is defined as

$$H_2(u, v) = \int_{-\infty}^{\infty} \int_{-\infty}^{\infty} h_2(x, y) \exp\{-2\pi i(ux + vy)\} \, dx \, dy \tag{4}$$

where (x, y) and (u, v) are spatial and spatial-frequency coordinates, respectively, and $h_2(x, y)$ is the PSF, which is assumed to be normalized; namely,

$$\int_{-\infty}^{\infty} \int_{-\infty}^{\infty} h_2(x, y) \, dx \, dy = 1 \tag{5}$$

The OTF is generally a complex quantity and can be described by its modulus $|H_2(u, v)|$ and its phase, $\phi_2(u, v)$, which are called the modulation transfer function (MTF) and the phase transfer function (PTF), respectively. The OTF can be expressed as

$$H_2(u, v) = |H_2(u, v)| \exp\{i\phi_2(u, v)\} \tag{6}$$

When the response of a radiographic imaging system, such as a screen-film system, is rotationally symmetric, the expression for the OTF can be simplified. In such a case, the LSF $h(x)$ is related to the MTF $H(u)$ by the Fourier cosine transform, namely,

$$H(u) = 2 \int_{0}^{\infty} h(x) \cos 2\pi ux \, dx \tag{7}$$

and these simplified one-dimensional quantities can be obtained from the two-dimensional quantities in the following manner:

$$H(u) = |H_2(u, 0)| \exp\{i\phi_2(u, 0)\} \tag{8}$$

and

$$h(x) = \int_{-\infty}^{\infty} h_2(x, y) \, dy \tag{9}$$

The phase term disappears in many cases when the LSF is symmetric.

Even if the PSF is not rotationally symmetric, the one-dimensional approach is commonly employed for experimentally determining the MTF (and PTF) because of the simplicity of the instrumentation and the procedure, as described in detail in Section 3.2. In this case, however, it is often necessary to determine the LSFs and MTFs in various directions; they can be derived by changing the orientation of the slit that is used as a line object.

The usefulness of the OTF and MTF is due to the fact that they are the most accurate and sophisticated tools presently available for image analysis. The OTF has been widely used in evaluating imaging components and also the imaging system as a whole. The OTF is particularly useful for a cascaded imaging system, since the overall OTF is simply given by the product of the OTFs of the components, whereas the convolution integrals are required in the spatial domain.[27] Many approaches to image processing[28,29] and analyzing image noise[30,31] have been evaluated by means of the OTF as well.

3.1. Methods of Measuring MTFs

The OTFs or MTFs of radiographic imaging components, such as screen-film systems and X-ray tube focal spots, have been measured by many different techniques. Basically, they may be divided into two approaches: (1) the Fourier transform of the LSF or (2) measuring the contrast of a cyclic pattern.

In the first approach, based on Fourier analysis, the LSF or the slit image is obtained experimentally, and then the digital Fourier transform is performed by means of a computer.[32,33] Analog techniques for the Fourier transform have also been attempted with a coherent optical system that requires a pinhole or slit image on a photographic transparency,[34] or with an incoherent optical system that employs a sinusoidal mask[35] or a Moiré pattern[36] for an emitted slit image (light distribution). When computers became readily available, the digital approach became popular for determining MTFs. Details of MTF measurements for screen-film systems and focal spots by digital Fourier transformation are described in Sections 3.2 and 3.3.

The LSF may be determined directly from the slit image as previously described, or it may be calculated from the first derivative of the edge response.[37] One problem associated with the edge-response method is that the first derivative is very sensitive to fluctuations in experimental data. In addition, it is difficult to determine accurately the tail portion of the LSF. Both of these factors may cause uncertainties in the calculated LSF and thus in the MTF as well.

In the second approach, based on contrast measurements, a cyclic pattern, such as a sinusoidal or square-wave pattern of varying frequency, is used as an object, and the contrast of the resulting image is determined experimentally. The ratio of the image contrast to the object contrast, plotted against the spatial frequency, provides the MTF if the sinusoidal pattern is

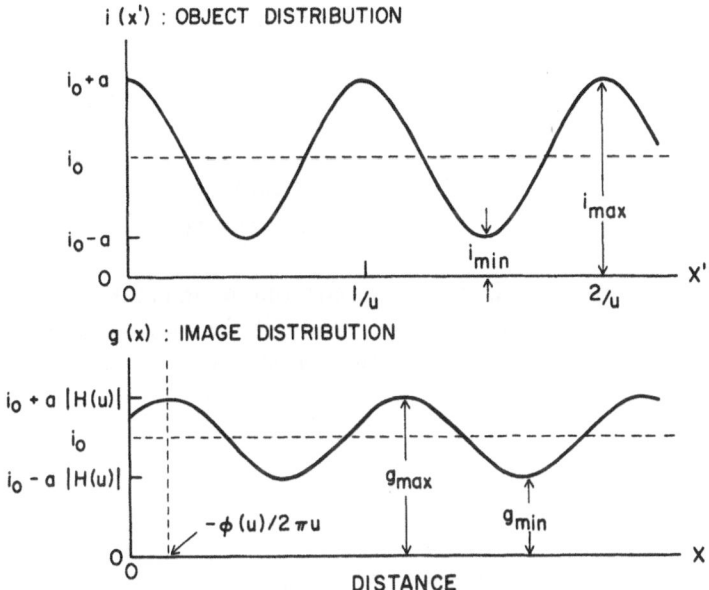

Figure 7. Relative intensity distributions of a sinusoidal object and the corresponding image in a linear and shift-invariant system.

employed or the square-wave response if the square-wave pattern is used. The MTF can be derived from the square-wave response as described later.

The principle of determining the OTF or MTF by the contrast method can be described as follows: We assume that the relative radiation intensity distribution of an object or the input to an imaging system is described by

$$i(x') = i_0 + a \cos 2\pi u x' \tag{10}$$

where i_0 is the average relative intensity, a is the amplitude of the sinusoidal pattern, u is the spatial frequency of this input distribution, and x' is the distance in the object plane, as illustrated in Figure 7. When this object is imaged by a linear and shift-invariant (or isoplanatic) system, such as an X-ray fluorescent screen, the image or output distribution $g(x)$, which is considered here as the relative light intensity distribution, is given by

$$g(x) = i_0 + a|H(u)| \cos\{2\pi u x + \phi(u)\} \tag{11}$$

as shown in Figure 7. Here, $|H(u)|$ and $\phi(u)$ are the MTF and PTF, respectively, and are defined in the same way as in Eqs. (4) and (6) in one-dimensional form

$$H(u) = \int_{-\infty}^{\infty} h(x) \exp(-2\pi i u x) \, dx$$

$$= |H(u)| \exp\{i\phi(u)\} \tag{12}$$

By using the Fourier cosine and sine transform, one can derive

$$|H(u)| = \left[\left\{ \int_{-\infty}^{\infty} h(x) \cos 2\pi ux \, dx \right\}^2 + \left\{ \int_{-\infty}^{\infty} h(x) \sin 2\pi ux \, dx \right\}^2 \right]^{1/2} \tag{13}$$

$$\tan \phi(u) = \frac{\int_{-\infty}^{\infty} h(x) \sin 2\pi ux \, dx}{\int_{-\infty}^{\infty} h(x) \cos 2\pi ux \, dx} \tag{14}$$

Equation (11) can be obtained from the convolution integral, which provides the fundamental relationship between the object, the image, and the LSF in a linear and shift-invariant imaging system, as shown by

$$g(x) = \int_{-\infty}^{\infty} i(x')h(x - x') \, dx'$$

or   (15)

$$g(x) = \int_{-\infty}^{\infty} i(x - x')h(x') \, dx'$$

The radiation contrast incident on a recording system was defined in Section 2.3 as $\Delta E/E$. We shall define the contrast of the input pattern in Eq. (10) in a similar way by applying the amplitude and average for ΔE and E so that the object contrast C_0 is equal to a/i_0. In practice, it is convenient to relate the object contrast to the maximum and minimum values of the cyclic pattern. One can show from Eq. (10) that

$$C_0 = \frac{a}{i_0} \tag{16}$$

or

$$C_0 = \frac{i_{max} - i_{min}}{i_{max} + i_{min}} \tag{17}$$

In a similar way, the image contrast C_i is defined in terms of the amplitude and average of the image distribution in Eq. (11) and also the maximum and minimum values of the image distribution; namely,

$$C_i = \frac{a|H(u)|}{i_0} \tag{18}$$

or

$$C_i = \frac{g_{max} - g_{min}}{g_{max} + g_{min}} \tag{19}$$

If we take the ratio of the image contrast to the object contrast in Eqs. (16) and (18), we obtain

$$\frac{C_i}{C_0} = |H(u)| \tag{20}$$

which is equal to the MTF. This is the basis for determining the MTF from measurements of the contrast in sinusoidal patterns.

Generating an X-ray intensity distribution with sinusoidal variation in space is very difficult, however. Hofert[5] attempted to produce approximately sinusoidal patterns to measure the MTF of screen-film systems by using an aluminum wedge filter and a special device that can move the filter and the cassette in such a way that the intensity of a narrow X-ray beam transmitted through the filter varies sinusoidally with time. The same technique was later applied in determining MTFs of X-ray films used with high-energy X-rays for industrial radiography.[38] In principle, the sine-wave response method has the advantage of obtaining the MTF from direct physical measurements. With these approaches, however, it is difficult to produce high-spatial-frequency patterns and also avoid a change in X-ray beam quality in the radiation intensity distribution.

The square-wave pattern has, therefore, been employed frequently to measure MTFs of screen-film systems[17,39-44] and focal spots.[45] The square-wave pattern can be produced rather easily if a lead test object[41] or a variable-slit device[6] is used. The square-wave response $H_{sq}(u)$, which can be determined from contrast measurements in the same way that it is used for the sine-wave response or the MTF, is related to the MTF $H(u)$ through Coltman's equations:[39]

$$H_{sq}(u) = \frac{4}{\pi} \sum_{k=1}^{\infty} (-1)^{k-1} \frac{H\{(2k-1)u\}}{(2k-1)} \tag{21}$$

$$H(u) = \frac{\pi}{4} \sum_{k=1}^{\infty} B_k \frac{H_{sq}\{(2k-1)u\}}{(2k-1)}$$

where

$$B_k = 0 \qquad\qquad m < n$$
$$= (-1)^n(-1)^{k-1} \qquad m = n \tag{22}$$

and n is the number of prime factors other than unity in $(2k-1)$ and m is the number of unique prime factors other than unity in $(2k-1)$. Therefore, the MTF can be calculated from the measured square-wave response. Commercially available lead test patterns consist of thin lead foils whose thickness ranges approximately from 100 to 200 μ. Lead foils are supported by sheets of plastic of approximately 2-mm total thickness. It has been found[44] that the scatter from the plastic part of the test pattern can influence the measured square-wave response; therefore, careful attention to the input X-ray pattern due to a given lead test object is necessary to accurately measure MTFs.

3.2. MTF of a Screen-Film System

The unsharpness or the LSF of a screen-film system is caused by processes that include light diffusion in the screen phosphor layers, crossover exposure between screens and films,[46-48] and cassettes that do not provide good contact between screens and films. The oblique incidence of X rays on a screen-film system produces a slightly shifted, double image on the front and back emulsions of the film. When the incident angle of the X rays is large, as, for example, in tomography, this effect may cause appreciable image unsharpness, which may be comparable to unsharpness due to light diffusion.[49-51] Recent studies[21, 52, 53] demonstrated that fluorescent X rays emitted by incident X rays having energies above the K absorption edge of the heavy elements in a screen phosphor may produce considerable scattering in the screens and thus cause a degree of unsharpness that may be comparable to the light diffusion inherent in the screens.

When transfer function analysis is applied to screen-film systems, it is usually assumed that the response of a screen-film system is linear, isoplanatic, and isotropic. The screen-film system is considered to be linearized when the density distribution is converted to the corresponding *effective* exposure (or illuminance) distribution by using the H and D curve. The word *illuminescence* has been used frequently for radiographic screen-film systems because the linearization procedure was first employed in photography. The screen-film system has been regarded as isoplanatic and isotropic because the manufacturing processes for screens and films are assumed to provide uniform products. Figure 8 shows images of three screen-film systems exposed to a narrow beam of X rays that is obtained with a

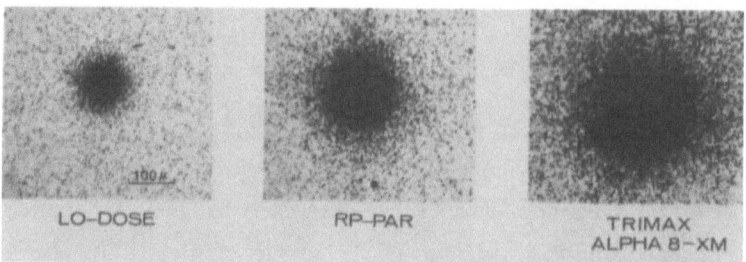

Figure 8. Pinhole images of screen-film systems demonstrating the isotropic nature of the point spread functions.

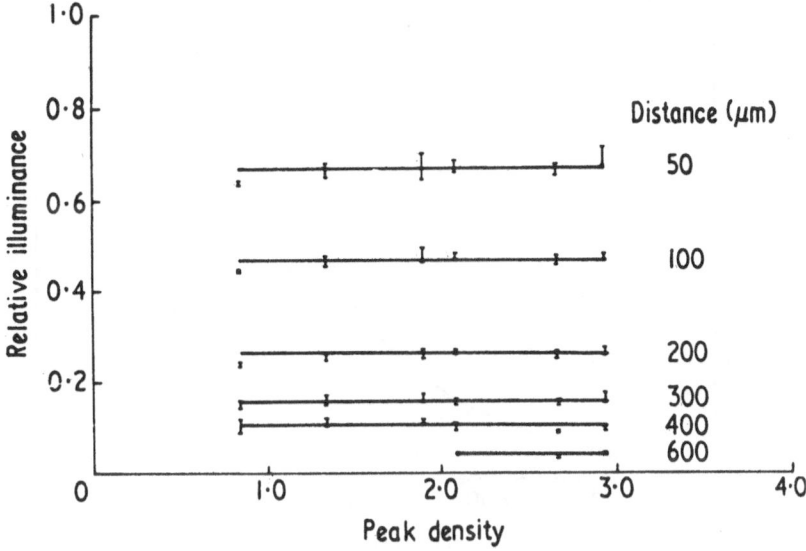

Figure 9. Density dependence of the LSF of a fast screen-film system.

12 μm × 12 μm platinum aperture. These "pinhole" images indicate the isotropic nature of the point spread functions of these screen-film systems.

Macroscopic nonlinearities in radiographic systems can be eliminated by means of their H and D curves. However, there are other factors that may affect microscopic nonlinearities in screen-film systems, namely, double-coated film; forming a direct X-ray image; and photographic nonlinearity, such as edge enhancement due to the development effect. The nonlinearity caused by double-coated film has been investigated both theoretically and experimentally,[54-56] and the conclusion was that, under the usual experimental conditions, the error introduced by this effect is about the same or less than the experimental error. However, under unusual conditions, such as an extremely asymmetric combination of front and back screens, the error due to nonlinearity can exceed the experimental error and influence the accuracy of the MTF measurement. The effects of the other two factors on the linearity of screen-film systems are less well known and usually neglected.

If these nonlinear factors influence MTF and LSF measurements, it is expected that the measured quantities depend on the density of the images to be analyzed. Figure 9 shows the density dependence of the LSF of a fast screen-film system.[57] The relative illuminance at various distances is plotted against the peak densities of slit images. There is no appreciable trend in the dependence of the LSF on density; therefore, this screen-film system is considered to be a linear system. Another way of examining the linearity of a screen-film system is demonstrated in Figure 10, where the measured edge

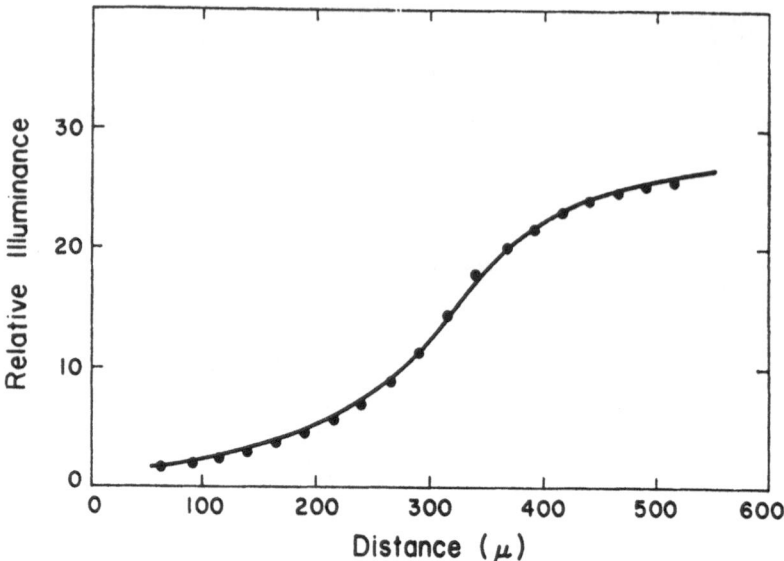

Figure 10. Comparison of measured edge response (dots) with calculated result (solid curve) for a medium-speed screen-film system (K. Rossmann, 1964).

response (dots) is compared with the solid curve calculated from the LSF.[58] The two results are in good agreement, which indicates the consistency and accuracy of the measurements and provides a basis for the assumed linearity of the screen-film system.

3.2.1. MTF MEASUREMENT BY THE SLIT METHOD

In our laboratory, the slit method and the digital Fourier transformation[33] are currently employed for MTF measurements. The main reasons for this choice are the following:

1. Both LSF and MTF are measured; the LSF is, of course, obtained by direct measurement.
2. The instrumentation is rather simple; aligning the slit is easier than aligning the edge.
3. Extensive information on various technical factors affecting this measurement is now available.

A block diagram for MTF measurements by the slit method is shown in Figure 11. In this method, two separate slit exposures are made with the screen-film system: a low-intensity exposure to determine the central portion of the LSF and a high-intensity exposure to evaluate the tails. The exposed

MTF MEASUREMENT BY SLIT METHOD

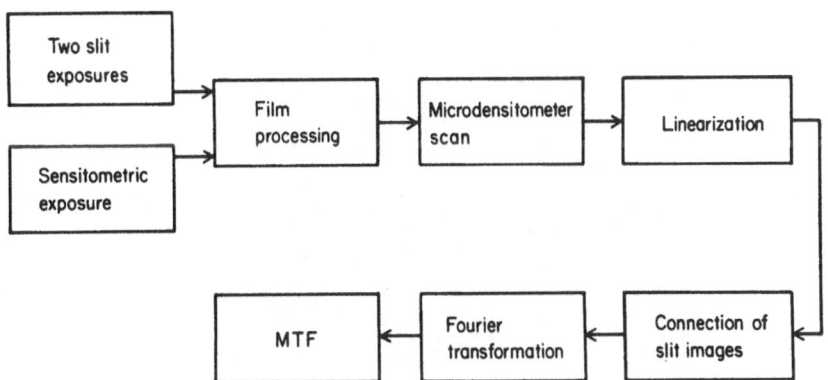

Figure 11. Block diagram of MTF measurement of screen-film system using slit images and digital Fourier transform.

film is developed together with a sensitometric strip that is obtained from an X-ray intensity scale, inverse-square sensitometer. The slit images and the sensitometric strip are scanned by a microdensitometer. The LSF is then obtained by converting the density to the relative illuminance by means of the H and D curve and by connecting two partial slit images. The MTF is then derived from a digital Fourier transformation of the LSF.

Figure 12 shows a photograph of the slit device. The slit, made of a 2-mm-thick platinum alloy, is aligned with the X-ray beam. The screen-film system is placed in vacuum contact with the slit, which eliminates not only poor contact between screens and film, but also the effect of geometric unsharpness. The slit device is aligned with the X-ray beam by either being rotated or shifted in a direction perpendicular to the X-ray beam. Figure 13 shows projected slit images at approximately 130 cm behind the slit; these images were obtained by the latter method at several positions on the slit device. For this demonstration, the position of the slit device was varied in 0.5-mm steps. Misalignment can be detected easily by comparing the darkness, narrowing, and symmetry of these projected slit images. In our laboratory, we have achieved precise alignment by comparing a similar series of slit images made in steps of 0.1 mm.

3.2.2. FACTORS AFFECTING MTF MEASUREMENTS

Technical factors that affect MTF measurements by the slit method include the width of the exposing slit, the truncation effect, and the sampling distance of the LSF data. Choosing the appropriate slit width is important because it can degrade the slit image and introduce a direct X-ray image of

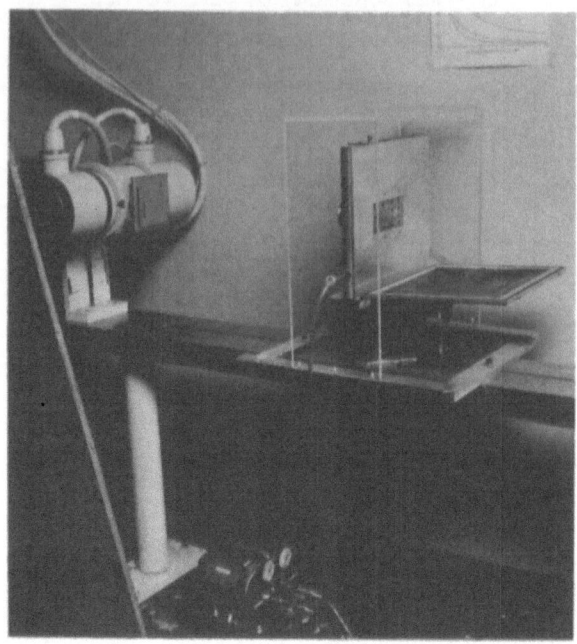

Figure 12. Photograph of the slit device for obtaining slit images on screen-film systems.

the slit. The narrower the slit width, the smaller the degradation of the LSF and the MTF will be, but the more the direct X-ray image appears. Based on theoretical analysis of this degradation and on experimental results for the direct X-ray image,[52] we are currently employing a 5- to 10-μ slit, which has a negligible effect on MTF measurements.

Although the LSF may have an infinite spatial extent, it is usually measured to approximately 0.01 of its maximum value for experimental reasons. This, in effect, truncates the LSF and introduces oscillating

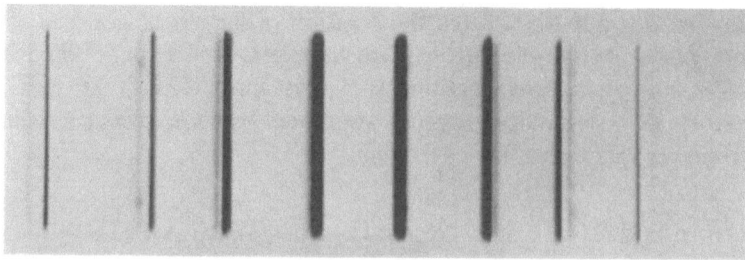

Figure 13. Projected slit images, obtained at several positions of the slit device, demonstrating the alignment of the device with the X-ray beam.

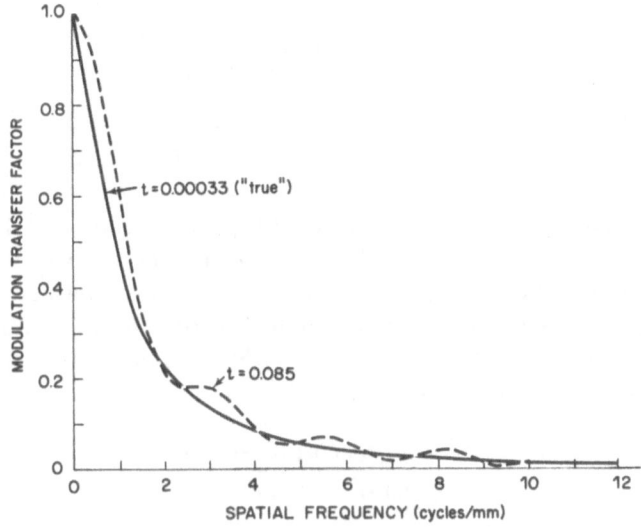

Figure 14. MTFs calculated from truncated LSFs for a fast screen-film system; t is the ratio of the level of LSF at the point of truncation to its maximum.

perturbations in the MTF, thus limiting its accuracy.[59,60] This error due to the lack of data for the tail portion of the LSF is called the truncation error. It has been shown[60] that the magnitude of the truncation error is related to the area of the tail portion of the LSF; that is, the portion that is usually not measured. This indicates that a significant error may enter into computing the MTF from an experimentally obtained LSF. Even if the truncated level of the LSF is small, it is possible that the area under the unknown tail portion is large and, therefore, results in a considerable error in the MTF. Figure 14 illustrates the comparison of MTFs computed from LSFs truncated at two significantly different levels.

In order to reduce the truncation error, we employed two techniques, namely, multiple slit exposure and exponential extrapolation of the LSF tail.[60] With the multiple-exposure technique, two separate slit exposures are made, and the LSF can then be measured to approximately 0.01 of its maximum value. The remaining truncation error, which can be considerable at low spatial frequencies, is then corrected by exponential extrapolation of the long sweeping tail.

If the tail LSF is given by the exponential approximation

$$h_T(x) = a \exp(-|x|/b) \qquad |x| > d$$
$$= 0 \qquad\qquad\qquad |x| \leqslant d \qquad (23)$$

then the Fourier transform is derived as

$$H_T(u) = 2ab \exp(-d/b) \frac{\cos(2\pi\, du + \phi)}{[1 + (2\pi bu)^2]^{1/2}} \qquad (24)$$

where $\phi = \tan^{-1}(2\pi bu)$. Since the tail LSF in Eq. (23) is extrapolated from the truncated LSF which is obtained experimentally, the constants a and b can be determined if one assumes that the exponential curve goes through the points (x_1, t_1) and (x_2, t_2) on the truncated LSF. The point (x_2, t_2) is the end point of the truncated LSF; that is, $x_2 = d$ and $t_2 = t$, where t_2 is the truncation level. The other point (x_1, t_1) is somewhat arbitrary; we determined experimentally that x_1 is equal to 3/4 or 5/6 of x_2. Substituting $b = (x_2 - x_1)/\ln(t_1/t_2)$ and $t_2 = a \exp(-x_2/b)$, the Fourier transform (F.T.) with the correction can be written as

$$\text{MTF} = \frac{[\text{F.T. of } (1/2)\text{LSF}] + [bt_2 \cos(2\pi x_2 u + \phi)]/[1 + (2\pi bu)^2]^{1/2}}{[\text{area of } (1/2)\text{LSF}] + bt_2} \qquad (25)$$

where the Fourier transform is applied to only one-half of the LSF because of the LSF symmetry.

Figure 15 shows the deviation of MTFs when they are computed from the measured LSF with and without the correction procedure. The truncation error without the correction can be as large as 0.07. Applying the

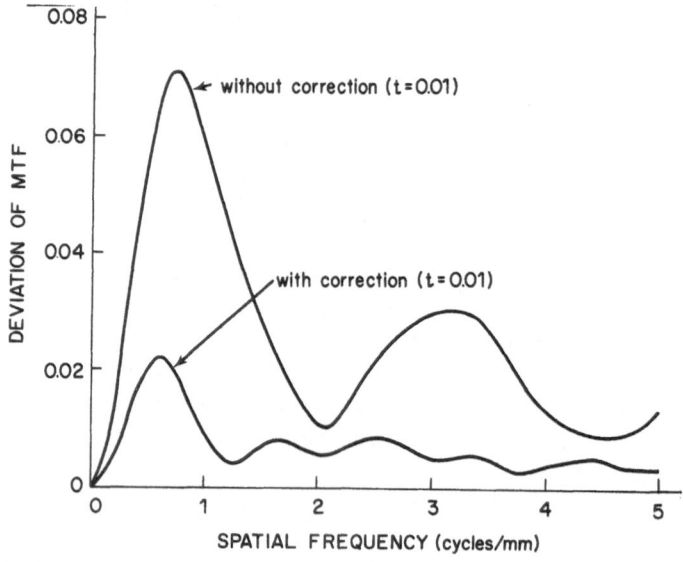

Figure 15. Comparison of truncation errors in computed MTFs of a slow screen-film system with and without correction.

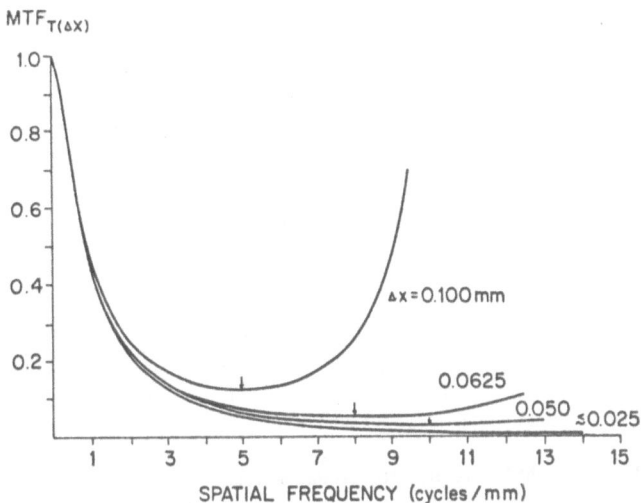

Figure 16. Demonstration of the effects of aliasing in MTFs computed from experimental LSF data of a fast screen-film system (HWHM; 0.085 mm) using the trapezoidal rule at several sampling distances.

correction technique reduces the truncation error to approximately 0.02, which is about equal to the precision of the experiment.

The aliasing error in the computed MTFs due to a finite sampling distance for the LSFs is illustrated in Figure 16. The aliasing error for screen-film systems generally occurs more often in the high- than in the low-spatial frequency range. The greater the sampling distance, the greater the aliasing error will be. Therefore, a simple method of reducing this error involves using the largest distance increment that gives a tolerable error in the computed MTF. Theoretical and experimental studies indicate that the sampling distance should be less than about 25% of the half-width (HWHM) of the LSF[61] if errors in the computed MTFs are to be less than 0.005.

3.2.3. EXPERIMENTAL RESULTS

The MTFs of four 3M Trimax screens combined with Kodak OG film, which were measured by the technique described in Section 3.2.1, are shown in Figure 17. These MTFs are averaged values obtained from two independent measurements. The reproducibility of the MTF is usually within 0.02 of the modulation transfer factor.[33] The uncertainty in the MTFs measured by the slit technique is usually greater at low than at high spatial frequencies, which may be related to the difficulty of determining the LSF tails accurately.

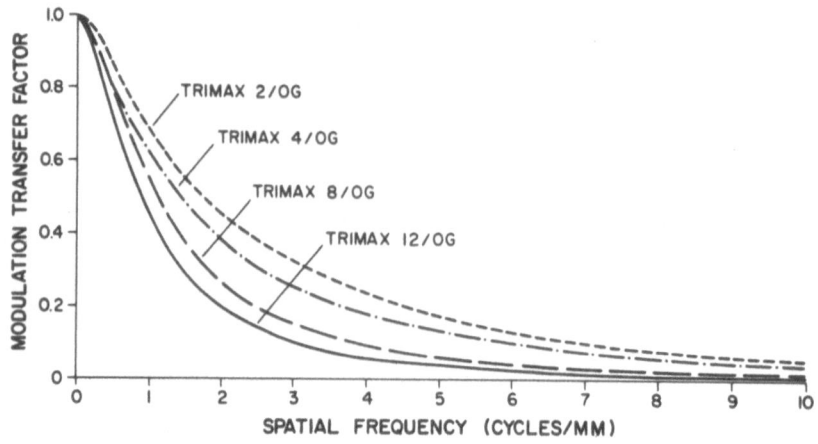

Figure 17. MTFs of four 3M Trimax screens with Kodak OG film.

In addition, the precision of determining the LSF and MTF of a slow, sharp screen-film system is generally lower than that for a fast, unsharp system perhaps because the slower the system, the narrower the LSF,[62] which makes the LSF measurement more difficult for the slow, sharp system.

The relative speeds of Trimax 2, 4, 8, and 12 screens with OG film are approximately 1.0, 1.5, 2.8, and 4.1, respectively. The slower the relative speeds, the greater the levels of the MTFs of these systems, as shown in Figure 17. That is, when the screen-film systems consist of the same film and of screens made of the same phosphor, resolution properties are usually related to the reciprocal of the speeds of these systems.

For many decades, it was believed that resolution properties of screen-film systems were determined mainly by the light diffusion in the screens and, therefore, they did not depend on the film used, but only on the screens. However, new technological developments in designing screens and films have made it clear that this concept must be changed.

Figure 18 shows the MTFs of Trimax 12 screens combined with four different films (Kodak OG and OM, 3M XD and XUD). Trimax 12/OM is a single-screen/single-emulsion film system;[51] a single Trimax 12 back screen is used. The XD film is a fast green-sensitive film, and XUD is an anticrossover film[48] that includes light-absorbing layers between the film emulsion and base. It is apparent that the MTFs of these systems differ significantly with the film used. The relative speeds of Trimax 12 screens with OM, XUD, OG, and XD film are approximately 2.1, 2.5, 4.1, and 8.4, respectively.

The MTFs obtained with OM and XUD film are considerably greater than those with OG and XD film because of the reduction of crossover

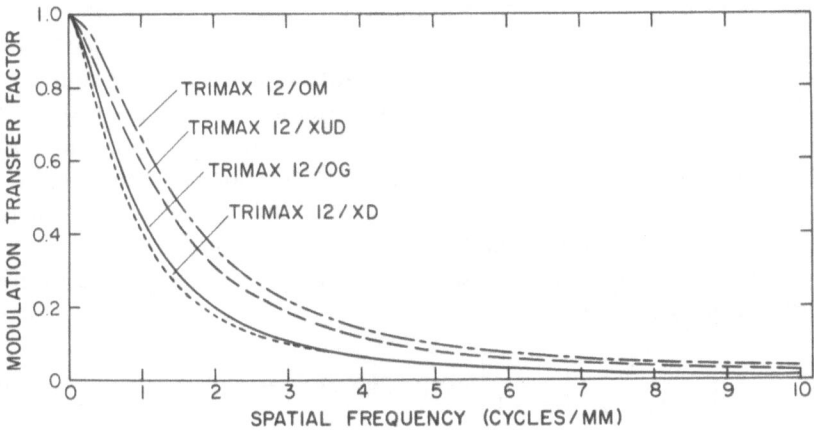

Figure 18. MTFs of Trimax 12 screens with four different films.

exposure in the screen-film systems. The MTF of Trimax 12/XD is slightly lower than that of Trimax 12/OG, which is also probably caused by the difference in crossover-exposure effects in these films that have significantly different speeds.

Recently, we observed degradation of the MTFs of blue-emitting $CaWO_4$ systems, including DuPont Detail, Par Speed, and Hi-Plus screens combined with XRP or DuPont Cronex 4 films, when we compared them with the MTFs of the corresponding systems obtained up to 1977.[33,63] We do not know the exact reason for this change; it may be related to a different crossover exposure arising from the recent use of films with a low silver content. This degradation in the MTF was approximately 0.03–0.07 in the spatial frequency range 1–2 cycles/mm. This difference in the MTFs was demonstrable in carefully prepared radiographs of test objects; its effect on clinical radiographs remains to be seen. However, these results indicate the need to carefully evaluate the MTFs of screen-film systems when a different film is used.

3.3. MTF of an X-Ray Tube Focal Spot

Radiographic images are essentially "shadow pictures." The X-ray intensity distribution, which is an input to the detector or image-recording system, such as a screen-film system and image intensifier, includes image degradation due to the finite size of the X-ray source; this effect has been called geometric unsharpness. The magnitude of geometric unsharpness is related to the X-ray intensity distribution $f(x)$ emanating from the source and the magnification factor m of an object projected onto the image plane. If the distance between the X-ray source and the object is d_1, and the distance between the object and the recording system is d_2, the image magnification factor m is $(d_1 + d_2)/d_1$. However, the magnification factor for the source size on the image plane, which equals the ratio of the source-to-object distance to the object-to-image distance, is $m - 1$. Therefore, the X-ray intensity distribution of the source at the image plane for a point object is $f\{x/(m-1)\}/(m-1)$, which corresponds to the LSF of a geometric unsharpness;[27, 64] it is normalized to have a unit integral over x.

The OTF of geometric unsharpness is therefore given by the Fourier transform of its LSF as

$$\int_{-\infty}^{\infty} \frac{1}{m-1} f\left\{\frac{1}{m-1}\right\} \exp(-2\pi iux)\, dx = F\{(m-1)u\} \qquad (26)$$

where $F(u)$ is the OTF of an X-ray tube focal spot, defined by

$$\int_{-\infty}^{\infty} f(x) \exp(-2\pi iux)\, dx = F(u) \qquad (27)$$

Equation (26) illustrates that the OTF or MTF of geometric unsharpness can be calculated from the OTF or MTF of the X-ray tube focal spot if one includes a factor $(m - 1)$ that simply shifts the spatial-frequency axis of the OTF of the focal spot. The factor $(m - 1)$ changes depending on the position of the object and on the focal spot-to-film distance. The OTF or MTF of the focal spot is, however, a unique imaging property of a given X-ray source distribution. Therefore, the OTFs or MTFs of the focal spots have often been determined for comparing imaging properties of the focal spots of X-ray tubes.

It is known that the shape and size of an X-ray tube focal-spot changes in a complicated way depending on a number of parameters, such as tube current and tube voltage,[34, 35, 65-76] as well as the direction along which the LSF or OTF is measured. In other words, the focal spot of a conventional X-ray tube is not rotationally symmetric. In addition, since the target plane of the X-ray source is not parallel to the image plane, the effective size and shape of the source distribution can vary significantly depending on the position in the object or image plane.[35, 71-79] These variations in geometric unsharpness over the image plane are called the field characteristics of the X-ray tube focal spot. It has been shown[79] that, under idealized geometric conditions, one can calculate field characteristics from the OTF at the central beam axis by taking into account the target angle and the field position.

The OTF of the focal spot usually contains a phase term due to the PTF at high frequencies, which often causes a distortion known as spurious resolution. The spurious resolution, which can be demonstrated easily if a test star pattern is used,[35, 80-82] can indicate the spatial frequency at the first minimum of the OTF. However, the significance of the PTF of the focal spot in clinical imaging is not well understood at present; the effect of the phase term may generally be negligible, since MTFs of other imaging components in a radiographic system decrease rapidly at high frequencies. Thus, in the final radiograph, there is almost no high-frequency, high-contrast object; such objects are most likely to be affected by the PTF. In fact, Burgess[83] has demonstrated that the effect of asymmetric focal spots, which yield the PTF, on blood vessel images was usually negligibly small. In the following section, therefore, we shall discuss primarily the MTFs of X-ray tube focal spots.

3.3.1. DEVICE FOR MEASUREMENT OF LSF AND MTF OF AN X-RAY TUBE FOCAL SPOT

The LSF of geometric unsharpness obtained at a magnification of two equals the LSF of the focal spot. Therefore, one can measure the OTF and other imaging parameters of the focal spot by placing an object halfway between the X-ray tube and a film and by recording the image on the film. For example, when a small pinhole is used as an object, one can obtain the PSF of the focal spot[34] and a slit for the LSF.[35,68,84,85] In addition, investigators have attempted to determine MTFs of X-ray tube focal spots by using a square-wave lead pattern[45] or a random test object.[86]

However, there are two difficult technical problems that are common to all methods, namely, aligning the test object or measuring device relative to the X-ray beam and localizing (or positioning) the focal spot. The latter, especially, may seriously degrade the accuracy of the measurement because geometric unsharpness is strongly dependent on the field position and the magnification factor of a test object placed in the X-ray beam.

We have developed a new device[85] with which accurate measurements of the focal spot distribution can be achieved. A photograph of the device, together with a diagnostic X-ray tube, is shown in Figure 19. The device contains two parallel support plates; the upper plate corresponds to the object plane and the lower, to the image plane. The top plate contains five pinholes of 1-mm diam. made of 1-mm thick lead disks. Four of the pinholes are located at the four corners of a 10-cm square, and the fifth is at the center of the square. The four peripheral pinholes are tilted in such a way that vertical lines through all five pinholes merge at a point 500 mm above the central pinhole. Thus, these multiple pinholes are focused at the point where the focal spot will be located. The bottom plate contains five X-ray fluorescent screens, each of which detects one of the magnified pinhole images. Four of the screens are located at the four corners of a 20-cm square, and one is at the center of the square. The center of each fluorescent screen is indicated by a red-light spot that is projected through a 200-μm pinhole placed beneath and in contact with the screen. The surface of the fluorescent

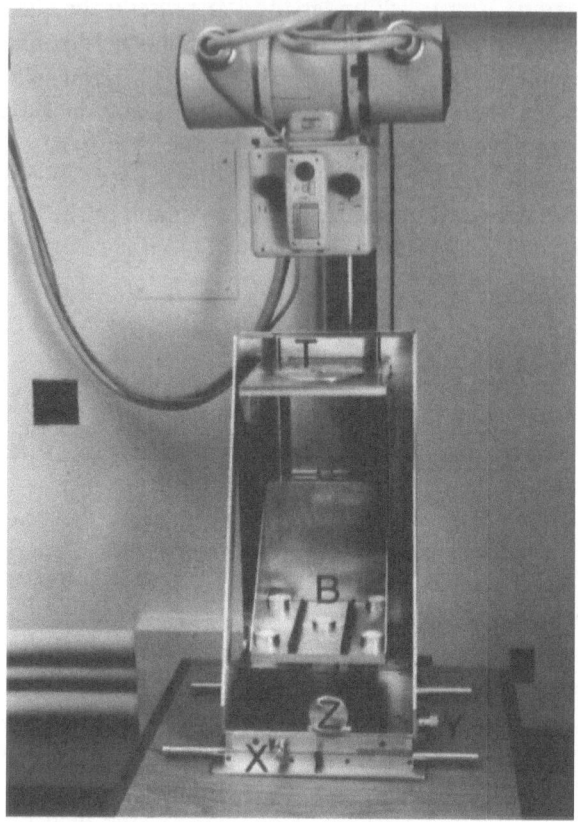

Figure 19. Photograph of a device for accurately measuring the X-ray tube focal spot. *T*: Top support plate containing five pinholes. *B̊*: Bottom support plate containing five X-ray fluorescent screens. A small vacuum cassette can be fitted into the bottom plate. *X, Y, Z*: Micrometer screws for adjustments, in three orthogonal directions, of the two support plates relative to the focal spot.

screen at the center is below the surfaces of the surrounding screens. A small vacuum cassette, not shown in Figure 19, fits above the central fluorescent screen on the bottom of the support plate. When the cassette is used, its surface is in a plane with the four surrounding fluorescent screens.

The distance between the two plates (the surface of the central pinhole disk and the surface of the four surrounding fluorescent screens) can be varied from approximately 300 to 500 mm by adjusting a micrometer screw behind the device. The standard distance employed for measuring the LSF and MTF of the focal spot is 500 mm. The position of these two plates relative to the focal spot can be varied in three orthogonal directions by three micrometer screws.

Operating this device is simple. First, it is placed below the X-ray tube at an appropriate position, and the X-ray tube housing and the device are leveled independently. Then the X-ray beam is turned on, and fluorescent images and red reference-light spots are viewed by a dark-adapted observer in a dark room. In this initial step, green spots usually do not coincide with the red spots at all five locations. Next, the position of the two support plates is varied in three directions relative to the focal spot until all pairs of green and red spots overlap. In this step, it is convenient first to use the green–red spots at the center for horizontal adjustments and then use the surrounding four pairs for vertical adjustment. However, it is often necessary to repeat the horizontal and vertical adjustments. The focal spot is now positioned on a line passing through the central pinhole and perpendicular to the two plates, at a distance from the top plate that is equal to the distance between the top and bottom plates. With this method, triangulation to locate the focal spot is unnecessary. The reproducibility of the adjustment was evaluated in terms of the standard deviation of micrometer readings at which all adjustments are completed. In our measurements, standard deviations for horizontal and vertical adjustments were 0.2 and 0.5 mm, respectively. These figures represent the precision with which the device can be positioned relative to the focal spot.

When this initial adjustment has been completed, one can make the following three types of measurements, with a known geometry, related to the focal spot as well as study the effect of the focal spot on radiologic imaging:

1. The pinhole image, or the PSF of the focal spot, can be obtained if the central pinhole is replaced with a small pinhole made of a Pt alloy. The size of the pinhole used depends on the size of the focal spot. A direct X-ray film or screen-film system, loaded in the small vacuum cassette that fits on the bottom support plate, can be used to record the pinhole image.

2. The slit image, or the LSF, can be obtained by replacing the multiple-pinhole insert with a precision slit device. The MTF is then calculated by digital Fourier transformation of the measured LSF.

3. Radiographs of test objects, such as a blood-vessel phantom, square-wave test object, and star pattern, can be made by means of another insert in the top support plate, which contains a 20 mm × 50 mm opening.

Figure 20 gives a close-up view of the precision slit device that replaces the multiple pinholes on the top plate. The slit device can be rotated to examine the anisotropic LSF and MTF of the focal spot. The slit is made of 1-mm-thick Pt alloy, and its width is variable up to 1 mm in increments of approximately 1 μm. The slit width is adjusted by a dual-thread method that is illustrated in Figure 21. The slit is composed of two jaws; one is fixed, and

Figure 20. Photograph of a precision slit device that here replaces the multiple pinholes on the top support plate.

the other is movable in dovetail fashion. The movable jaw is connected to one side of a dual-thread screw, which has 40 threads/in (15.74/cm) on that side. The base of the slit holds the other side of the screw, which has 36 threads (14.17/cm). The two threads are made to move in the same direction. Thus, one revolution moves the screw by 1/36 in relative to the base, and, at the same time, the movable jaw moves in the opposite direction by 1/40 in relative to the screw. Therefore, the combined motion of the movable jaw relative to the fixed jaw is $(1/36 - 1/40) = 1/360$ in $(71 \, \mu m)$ for one

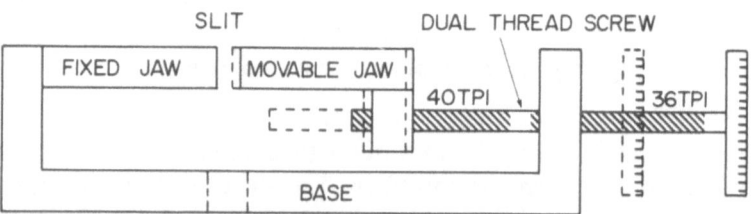

Figure 21. Illustration of the dual-thread method for fine adjustment of slit width. Dashed lines indicate that the movable jaw traverses one-tenth of the distance of the dual-thread screw relative to the base.

revolution of the screw. With this dual-thread method, the slit width is thus adjusted by a slow, smooth movement of the jaw. One of the advantages of this method is that the backlash is almost negligible compared to that in other methods in which gears and/or a tapered wedge are used. Details of the technical factors affecting measuring LSFs and MTFs with this device are discussed elsewhere.[85] Some results of measurements are described in the following section.

3.3.2. SIZE OF AN X-RAY TUBE FOCAL SPOT

There is no doubt at present that the LSF or MTF of the focal spot should be determined and should be used in rigorous scientific discussions of the evaluation of radiologic imaging, such as optimizing magnification techniques[87-92] and improving resolution properties in certain diagnostic examinations.[93,94] However, a single number has been used as an indicator of the resolution properties of an X-ray focal spot for many purposes—for example, as a part of specifications for an X-ray tube and for monitoring the focal spot in quality assurance programs.

Several methods for determining the size of an X-ray tube focal spot are presently available. Traditionally, the focal-spot size has been measured by the pinhole method.[95-99] During the last ten years, however, based on studies of imaging parameters for radiographic systems, investigators have proposed several methods using the star pattern,[80,96] slit,[99,100] sampling aperture,[33] root mean square (RMS) value,[101-103] and other factors.[104-107]

The advantage of using a single number rather than a curve, such as the MTF, is obvious; it is impossible, however, to represent a curve by a number in every case. The important question is which method provides the best estimate of the focal-spot size. Therefore, we carried out a study to demonstrate differences among various methods and show the correlation between the measured focal-spot size and corresponding image distributions in angiography.

Thirty-two focal spots of four X-ray tubes, used under various exposure conditions, were selected for this study. The LSFs and MTFs of these focal spots were measured, and focal-spot sizes were determined by the pinhole, star pattern, slit, and RMS methods. All of the measurements were made with the precision device described in Section 3.3.1. Therefore, focal-spot sizes and MTFs could be measured accurately in the center of an X-ray beam, which is defined here as a beam emanating from the focal spot in a direction perpendicular to the tube port. All test objects, namely, the pinhole, slit, and star pattern, were placed at the same position for measurements with this device.

3.3.2a. Pinhole Images. Pinhole images of large and small focal spots of the

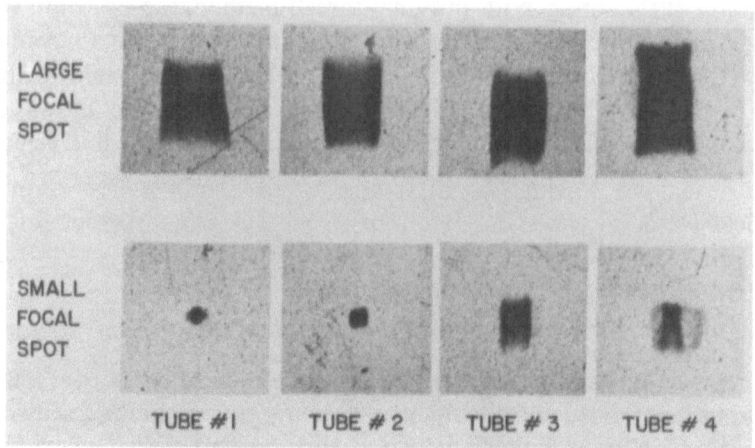

Figure 22. Pinhole images of large and small focal spots of the four X-ray tubes used in this study. The vertical axis in this figure corresponds to a direction parallel to the tube axis.

four X-ray tubes studied are shown in Figure 22. The X-ray generators and tubes used were two Siemens Tridoros 150G-3 units with Bi 125/3/50 RG (tubes no. 1 and no. 2), an Elema-Schonander Optimax 1024 with Philips Rotalex (tube no. 3), and a Philips Diagnostic 73 with Machlett Dynamax DX-69B (tube no. 4). Two generators and two tubes (no. 1 and no. 2) were the same models but had been purchased separately. Exposures were made at 75 kVp, with approximately one-half the maximum tube current allowed for each focal spot. The images were obtained with a 75-μm pinhole[95,96,108] for large focal spots and with a 30-μm pinhole for small focal spots, at 2× magnification and a 50-cm focal spot-to-film distance. Kodak XRP film was used without screens.

The same geometry and the same film were employed for other measurements as well; for example, Kodak dental X-ray film was also examined for use in focal-spot measurements. Dental X-ray film was approximately two times faster and less noisy than XRP but required manual processing. In our judgment, the difference in noise levels in XRP and dental X-ray films does not make a significant difference when measuring MTFs and focal-spot sizes. Thus, because of the convenience of machine processing, XRP film was used in this study. We obtained a series of pinhole images at one setting of the focal spot by changing the exposure time. Throughout this study, we varied only the exposure time to obtain radiographic images of the desired density. Pinhole images with comparable peak densities were selected subjectively for measuring focal-spot size. Focal-spot sizes were measured by five observers who used a 6× magnifier with a reticule having 0.1-mm increments. No correction was applied for differences in the shape of the focal spot in two directions.[95,109,110] The average standard deviation of measured focal-spot sizes among five observers was 0.03 mm for large focal spots and 0.04 mm for small focal spots. These results indicate that, although the pinhole method includes subjective judgment, variation in the focal-spot size as measured by trained observers is relatively small.

3.3.2b. *Slit Images.* To obtain slit images, we used a 50-μm slit for large focal spots and a 10-μm or 20-μm slit for small focal spots. The slit was placed perpendicular to the direction of the focal-spot measurements; i.e., the focal-spot size was measured in a direction parallel to the tube axis with a slit placed perpendicular to that axis. The peak density of the slit images was approximately 1.5. The average standard deviation of focal-spot sizes measured by the slit method was less than 0.03 mm. Slit images and X-ray sensitometric strips were processed together and used later for MTF measurements.

3.3.2c. *Star Pattern Images.* Focal-spot sizes were also determined with a two-degree star pattern at $2\times$ magnification for large focal spots. A one-degree pattern was used at $2\times$ magnification with 0.6-mm focal spots and at $3\times$ magnification with 0.2-mm focal spots. The *blur distance* of a star pattern is defined here as the mean diameter of the outermost distorted zone; this distance was measured with a ruler having 1-mm increments. The focal spot size f was calculated from the magnification m and blur distance d by the equation

$$f = \frac{kd}{m - 1} \tag{28}$$

where $k = 0.0347$ for a two-degree star pattern and 0.0174 for a one-degree star pattern.[80,96] The focal-spot size determined by the star pattern method generally corresponds to deriving an equivalent uniform focal-spot size that provides the same spatial frequency of the first minimum of the MTF as that of the measured focal spot. Advantages of the star pattern method are that images can be obtained with a relatively short exposure time and the measured focal-spot size usually does not depend on the density of the image. The standard deviation of effective focal-spot sizes obtained by the star method was 0.03 mm for large focal spots and 0.01 mm for small focal spots.

3.3.2d. *RMS Focal-Spot Size.* The focal-spot size determined by the RMS method has been called the RMS-equivalent uniform focal-spot size.[101] The RMS value is calculated from

$$\text{RMS} = \left[\int_{-\infty}^{\infty} x^2 g(x) \, dx \right]^{1/2} \tag{29}$$

where $g(x)$ is the LSF of the focal spot, which is normalized and centered in the following manner:

$$\int_{-\infty}^{\infty} g(x)\, dx = 1 \tag{30}$$

$$\int_{-\infty}^{\infty} xg(x)\, dx = 0 \tag{31}$$

The size of a uniform focal spot is equal to its RMS value multiplied by a factor of approximately 3.5; for simplicity, we shall call this the RMS focal-spot size. The advantages of this definition are (1) the total RMS value of a cascaded imaging system is equal to the square root of the squared sum of partial RMS values for all components; and (2) the RMS of the LSF, which is proportional to the curvature of the MTF at zero spatial frequency, is related directly to the low-frequency performance of the imaging system. The latter is important because the Fourier spectrum of the input X-ray pattern generally has more low-frequency than high-frequency content.

3.3.3. RESULTS OF FOCAL-SPOT MEASUREMENTS

Table 1 shows results obtained from four different methods of measuring large focal-spot sizes in two directions, i.e., parallel and perpendicular to the X-ray tube axis. Exposures were made at 75 kVp and approximately one-half the maximum tube current. There were slight variations in the measured focal-spot sizes. It appears that the size variation among different methods is greater in the parallel than in the perpendicular direction.

Table 1. Sizes (mm) of Large Focal Spots of Various X-Ray Tubes, Determined from the RMS Value of the Line Spread Function and from the Star Pattern, the Pinhole Image, and the Slit Image

Tube	Nominal focal-spot size	Exposure condition	RMS method	Star method	Pinhole method	Slit method
1[a]	1.0 mm	75 kVp, 250 mA	1.42	1.43	1.50	1.42
2[a]	1.0 mm	75 kVp, 250 mA	1.28	1.36	1.20	1.19
3[a]	1.0 mm	75 kVp, 250 mA	1.16	1.13	1.16	1.15
4[a]	1.2 mm	75 kVp, 500 mA	1.21	1.20	1.26	1.18
1[b]	1.0 mm	75 kVp, 250 mA	1.52	1.42	1.82	1.72
2[b]	1.0 mm	75 kVp, 250 mA	1.66	1.59	1.82	1.82
3[b]	1.0 mm	75 kVp, 250 mA	1.65	1.50	1.98	1.90
4[b]	1.2 mm	75 kVp, 500 mA	2.28	2.18	2.34	2.29

[a] Focal-spot size in a direction perpendicular to the tube axis.
[b] Focal-spot size in a direction parallel to the tube axis.

Table 2. Sizes (mm) of Small Focal Spots of Various X-Ray Tubes, Determined from the RMS Value of the Line Spread Function and from the Star Pattern, the Pinhole Image, and the Slit Image

Tube	Nominal focal-spot size	Exposure condition	RMS method	Star method	Pinhole method	Slit method
1[a]	0.2 mm	75 kVp, 40 mA	0.27	0.16	0.48	0.47
2[a]	0.2 mm	75 kVp, 40 mA	0.33	0.32	0.40	0.33
3[a]	0.6 mm	75 kVp, 130 mA	0.74	0.70	0.76	0.74
4[a]	0.6 mm	75 kVp, 100 mA	0.87	0.26	1.20	1.10
1[b]	0.2 mm	75 kVp, 40 mA	0.26	0.22	0.36	0.30
2[b]	0.2 mm	75 kVp, 40 mA	0.30	0.24	0.42	0.36
3[b]	0.6 mm	75 kVp, 130 mA	0.93	0.58	1.10	1.08
4[b]	0.6 mm	75 kVp, 100 mA	0.96	0.63	1.04	0.99

[a] Focal-spot size in a direction perpendicular to the tube axis.
[b] Focal-spot size in a direction parallel to the tube axis.

Measured sizes of four small focal spots are listed in Table 2, which shows that for small focal spots, the measured size depends significantly on the method used. For example, for tube no. 4 in the perpendicular direction, the pinhole method gave more than four times the focal-spot size obtained with the star method, and the RMS focal-spot size was approximately three times that obtained with the star pattern. Similar measurements were made for both small and large focal spots in X-ray tube no. 1 under four different exposure conditions, i.e., at high- and low-kVp settings and with high and low tube currents. The results are not shown here.

All focal-spot sizes measured are plotted in Figures 23–25, which demonstrate the correlation between focal-spot sizes determined by the different methods. Figure 23, which illustrates the correlation between the RMS and pinhole methods, shows that the pinhole method generally provides slightly larger focal-spot sizes than does the RMS method. Figure 24 shows a similar trend; the slit method also generally gives a larger focal-spot size than the RMS method. However, it is apparent from Figure 25 that the star method provides focal-spot sizes that are generally slightly lower than those for the RMS method. For a few of the focal spots examined, the star method resulted in significantly smaller focal-spot sizes.

3.3.4. CORRELATION BETWEEN MEASURED FOCAL-SPOT SIZES AND BLOOD-VESSEL IMAGES

In order to study the effect of measured focal-spot sizes on the quality of radiographic images, we calculated the image distributions of blood vessels in

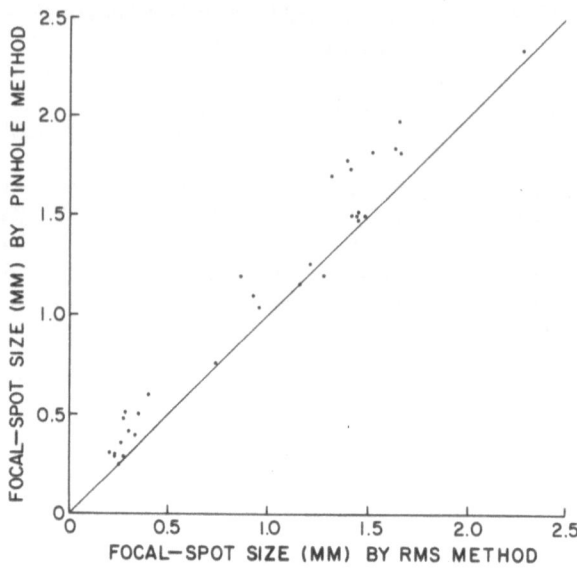

Figure 23. Correlation of focal-spot sizes determined by pinhole and RMS methods.

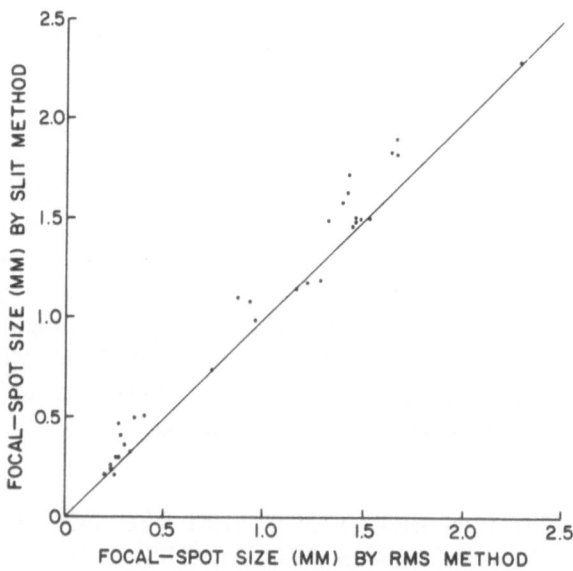

Figure 24. Correlation of focal-spot sizes determined by slit and RMS methods.

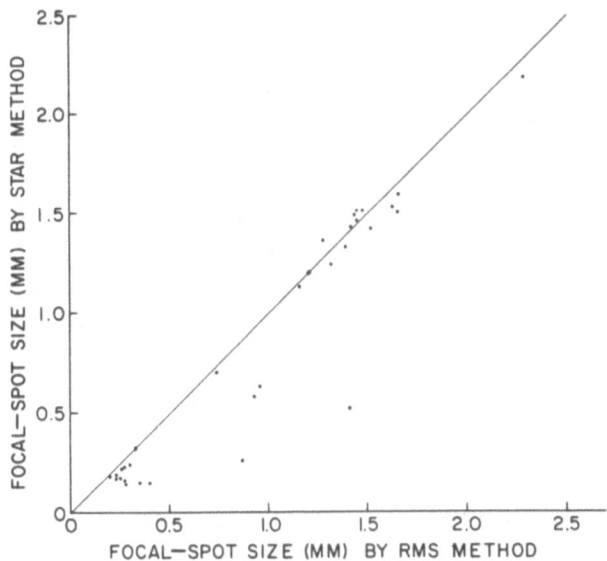

Figure 25. Correlation of focal-spot sizes determined by star and RMS methods.

angiograms by using a computer simulation technique.[111-114] This technique is particularly useful when the number of technical parameters involved is so large, as in angiography, that it is difficult to perform accurate experimental studies. It has been shown that the image distributions predicted by a computer agree with the experimental results.[115] Figure 26 shows the geometric arrangements used for two groups of focal-spot sizes. The standard technique illustrated on the left is applied to large focal spots used with $1.2 \times$ magnification and a medium-speed screen-film system. The diameter of the blood-vessel phantom was varied from 0.1 to 1 mm, and the linear attenuation coefficient was derived by exposing the contrast medium (Renografin-60) at 80 kVp.[115] The image distribution of a blood vessel filled with the contrast medium was calculated by two convolutions of the input X-ray pattern of the vessel with the LSFs of geometric unsharpness and the screen-film system (DuPont Par Speed screens with Kodak XRP film). The LSFs of geometric unsharpness that we used were the true LSFs obtained from direct measurements as well as LSFs of uniform shape having focal-spot sizes determined by measurements based on different techniques. When the focal-spot size is described by only a single number, the shape of the LSF is assumed to be uniform, with its width being equal to that number.

The geometric arrangement on the right was applied to small focal spots and calculating vessel image distributions at $2 \times$ magnification, with a high-speed screen-film system (GAF TF-2 screens with Kodak XRP film).

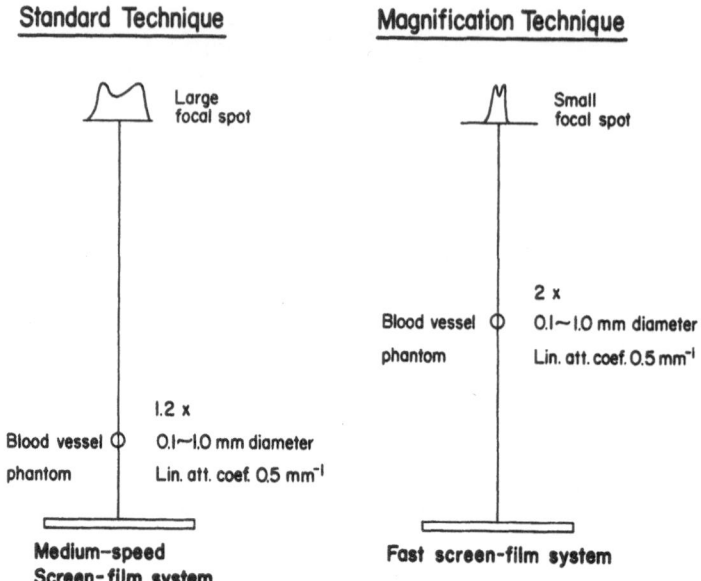

Figure 26. Geometries and parameters used to calculate image distributions of blood vessels for determining the effect of measured focal-spot sizes on radiologic images.

Figure 27 illustrates image distributions of a 0.4-mm vessel, obtained with various focal-spot sizes determined by the pinhole, star, slit, and RMS methods, and also the *true* image distribution computed from the focal-spot LSF in a perpendicular direction for X-ray tube no. 1. There was no difference in the image distributions (i.e., the curves coincided exactly), indicating that, for this particular focal spot and geometry, the methods used for focal-spot measurements do not affect the predicted quality of the vessel images.

However, a complicated variation in the image distribution of a 0.4-mm blood vessel at $2 \times$ magnification can occur, as demonstrated in Figure 28 for tube no. 4 in a perpendicular direction. It is obvious that the vessel image distributions obtained from both pinhole and slit focal spots are much broader and more unsharp than the true distribution derived from the focal-spot LSF; they also result in low contrast. The star method provided a much sharper distribution and higher contrast than did the true distribution. However, there is a close resemblance between the RMS-based distribution and the true image distribution. These results indicate that for this specific case, both the pinhole and the slit methods overestimated the focal-spot size, while the star method gave low estimates, and the RMS method provided a good estimate of the actual size of this focal spot.

We calculated similar image distributions of various blood vessels for 32

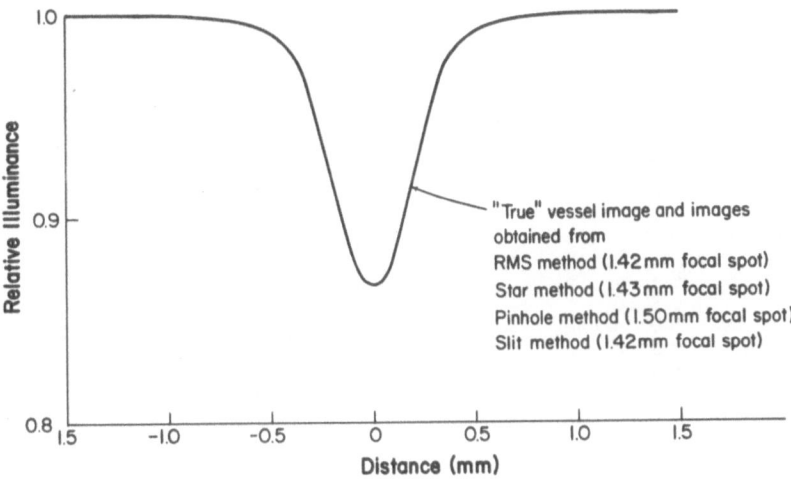

Figure 27. Image distribution of 0.4-mm blood vessel phantom at 1.2 × magnification with the large focal spot of tube no. 1 and with a medium-speed screen-film system.

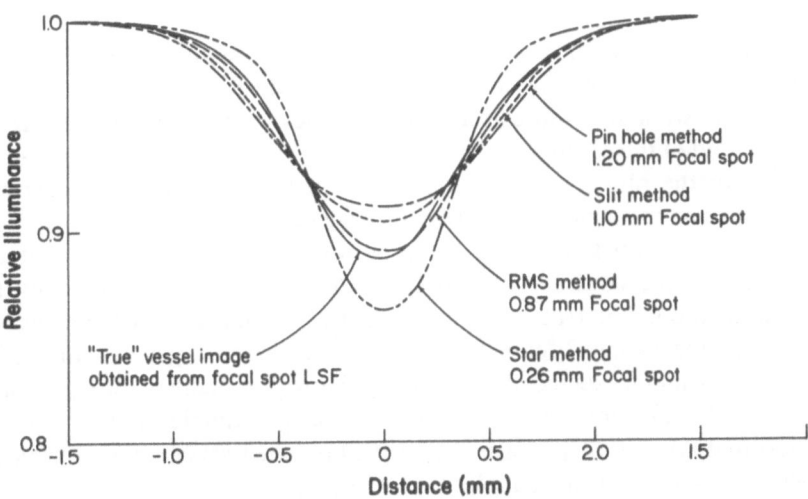

Figure 28. Image distributions of 0.4-mm blood vessel phantom at 2 × magnification with the small focal spot of tube no. 4 and with a fast screen-film system. Vessel images were obtained with measured focal-spot LSF and also uniform LSFs having focal-spot sizes measured by various methods.

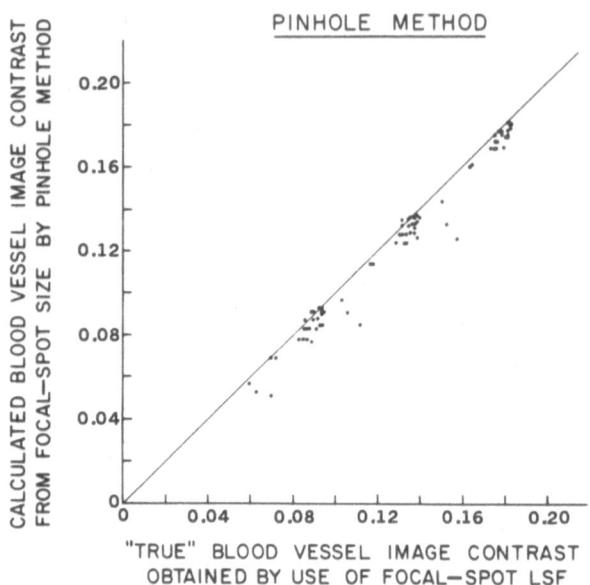

Figure 29. Relationship of true image contrast of blood vessels to predicted contrast based on focal-spot sizes measured by the pinhole method.

different focal spots. The results showed a considerable variation in the computed vessel image distributions; examples illustrated in Figures 27 and 28 indicate two somewhat extreme cases, however. In order to examine the overall trend of the results, we used a single figure of merit, namely, the image contrast of the blood vessels, which is defined as the relative decrease in illuminance at the center of the vessel image distribution.[111,115]

Figure 29 shows the relationship between the true blood-vessel image contrast and the contrast derived from focal-spot sizes measured by the pinhole method. The true image contrast was obtained from the measured LSFs of all focal spots. Figure 29 illustrates that there is a general correlation between true and predicted image contrast. However, many of the data points lie below the 45° *ideal* correlation line. Thus, the pinhole method tends to overestimate the focal-spot size so that the predicted vessel image contrast is likely to be less than the actual image contrast.

Figure 30 shows a similar relationship for vessel image contrast obtained by the slit method. From Figure 30, it appears that the image contrast obtained by the slit method is lower than the true contrast. The result for the star pattern method is illustrated in Figure 31. The image contrast predicted from the star method seems to be higher than the true contrast in some cases.

Figure 32 shows the correlation between the image contrast obtained by

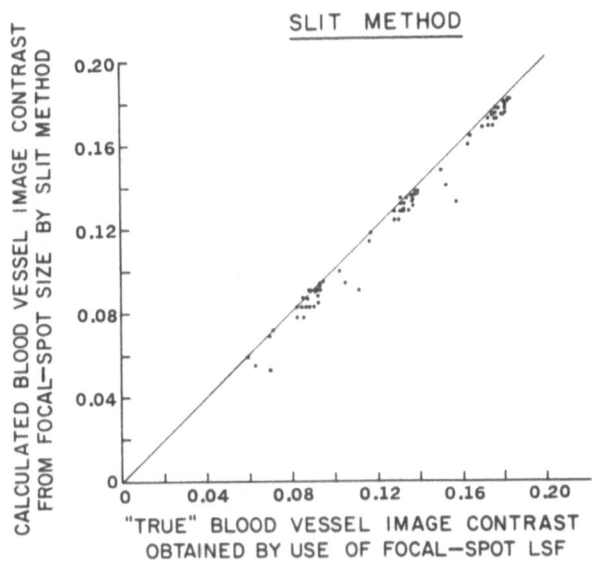

Figure 30. Relationship of true image contrast of blood vessels to predicted contrast based on focal-spot sizes measured by the slit method.

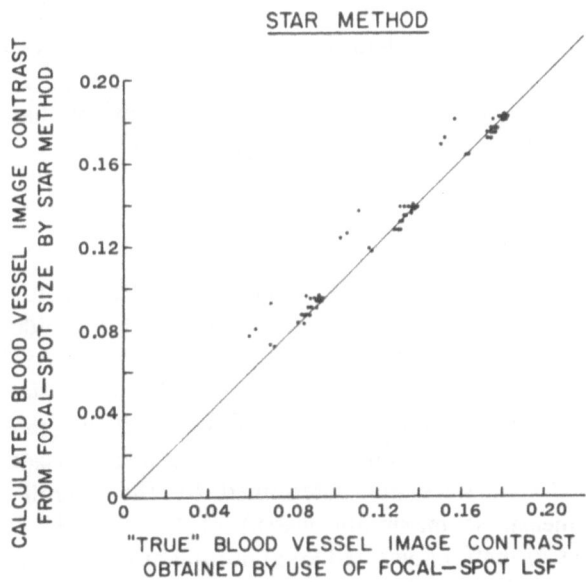

Figure 31. Relationship of true image contrast of blood vessels to predicted contrast based on focal-spot sizes measured by the star pattern method.

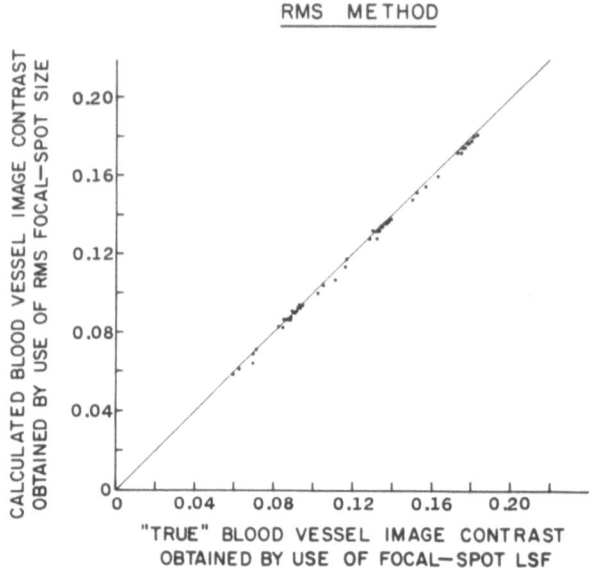

Figure 32. Relationship of true image contrast of blood vessels to predicted contrast based on focal-spot sizes measured by the RMS method.

the RMS method and the true image contrast. The correlation demonstrated in Figure 32 is much better than that for other methods. This indicates that focal-spot sizes determined by the RMS method are much more closely related to the quality of blood-vessel image distributions than are the sizes obtained by other methods investigated in this study.

3.3.5. LSFs AND MTFs OF FOCAL SPOTS

Figure 33 illustrates the LSFs and MTFs of small focal spots in a direction perpendicular to the tube axis; Figure 33 also includes two cases in which a large variation in measured focal-spot sizes occurred. The LSF of a 0.6-mm focal spot (tube no. 4) contains four peaks; focal-spot sizes determined by both pinhole and slit methods included all of these peaks. This resulted in relatively large values, as shown in Table 2, because, with these methods, focal-spot sizes are determined by the overall size (or the subjectively measured maximum width) of the pinhole or slit image. However, the star method detected an abnormal appearance of the image at high frequencies, which is related to the two sharp peaks near the center of the LSF and thus gave a small focal-spot size. The RMS method, however, provided an intermediate size for this focal spot. It should be noted that one

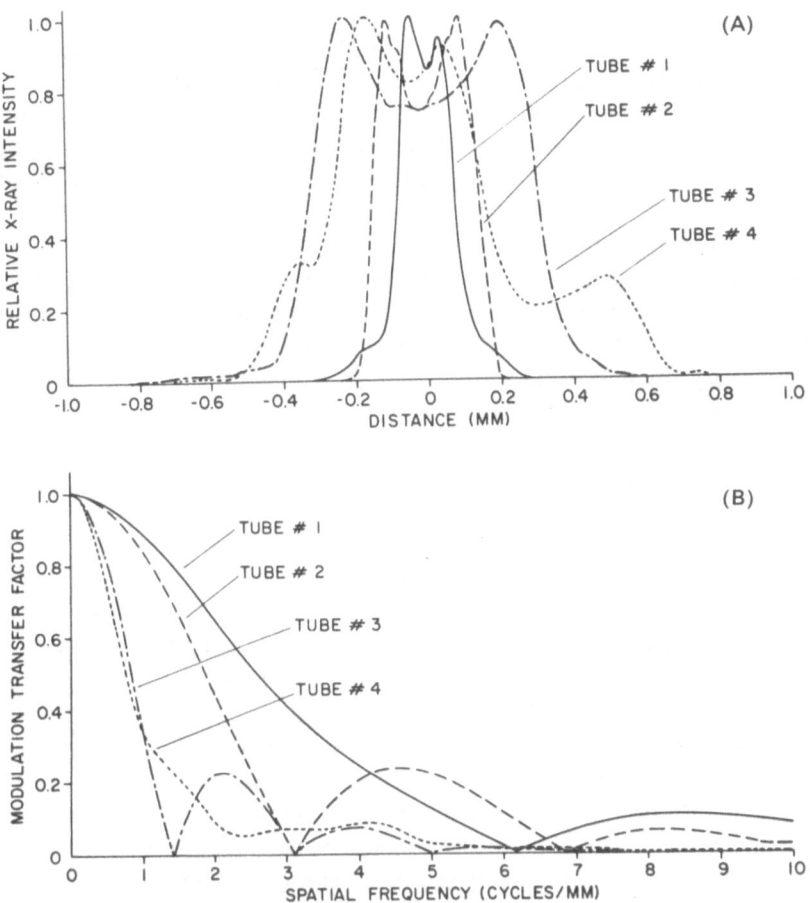

Figure 33. (A) LSFs and (B) MTFs of small focal spots in perpendicular direction, measured by placing a slit in a direction parallel to the tube axis.

determines the RMS value of the LSF by taking into account the overall distribution of the LSF with an appropriate weight, as described in Eq. (29). A measured focal-spot size very similar to that for tube no. 4 was obtained for tube no. 1, as shown in Table 2.

The MTF of the 0.6-mm focal spot (tube no. 4), which contained four peaks in the LSF, appears to have two or more components, as demonstrated in Figure 33(b). The star method detected two minima at high frequencies, located at approximately 2.4 and 5.1 cycles/mm; two observers detected 2.4 cycles/mm and three others, 5.1 cycles/mm, as the first minimum. These results were then averaged and thus provided a mean disappearance frequency of approximately 3.8 cycles/mm and a mean equivalent focal-spot size of 0.26 mm. It is obvious that the low-frequency performance of this focal spot is much lower than that indicated by this disappearance frequency. This example illustrates potential problems inherent in the star method of determining the focal-spot size; namely, (1) the disappearance frequency detected may be quite high compared to the dominant low-frequency performance; and (2) when the MTF of the focal spot contains multiple minima and the MTF in that frequency range changes gradually, as demonstrated for tube no. 4 in Figure 33(b), it is very difficult for a human observer to determine a unique point consistently. In practice, this has important implications. For example, if the focal spot is designed to produce one or more sharp peaks, the focal-spot size determined by the star method will be very small, but the actual imaging properties will be very poor.

The small focal-spot size of tube no. 3 in a direction perpendicular to the tube axis (Table 2) did not change much with the method used. The LSF of this focal spot, as shown in Figure 33(a), had a relatively smooth twin-peak distribution, and the approximate "width" of the LSF was considerably larger than the approximate "width" of the central part of the LSF of tube no. 4. However, the MTF of tube no. 3 was higher than that of tube no. 4 at low spatial frequencies. The higher MTF of tube no. 3 at low spatial frequencies is reflected in the measured RMS focal-spot size; i.e., the RMS focal-spot size of tube no. 3 in a perpendicular direction was less than that of tube no. 4.

For small focal spots, the LSF of tube no. 1 in the perpendicular direction contained a narrower central part and broader tails than did the LSF of tube no. 2, as shown in Figure 33(a). The focal-spot size of tube no. 1 determined by both pinhole and slit methods (Table 2) was larger than that of tube no. 2; this result is obviously affected by the broad tails. The focal-spot size of tube no. 1 determined by the star method was, however, considerably smaller (by a factor of two) than that of tube no. 2, due to the narrow central part of the LSF of tube no. 1. It should be noted, however, that the MTF of tube no. 1 at low spatial frequencies was improved by less than a factor of two. The focal-spot size of tube no. 1 determined by the RMS method was approximately 20% less than that of tube no. 2; this seems a rather reasonable difference between the two focal spots in view of the low-frequency performance of the two focal spots.

3.3.6. APPROXIMATING RMS FOCAL-SPOT SIZE

An important question that we wanted to answer in studies concerning focal-spot-size measurements was which method would provide the best estimate of the imaging properties of the focal spot in terms of a single number. Our results clearly indicated that the RMS method is preferable in this respect to the pinhole, slit, or star methods. However, another question is whether the RMS method is a practical one. The answer depends on many factors;[116-119] for example, a well-equipped laboratory will have few problems determining the RMS focal-spot size from the measured LSF. For many hospitals, however, it would be very difficult to obtain the LSF of an X-ray tube focal spot and, thus, the RMS focal-spot size. We found that it is possible to obtain a first-order approximation of the RMS focal-spot size from the arithmetic average of the slit and star focal-spot sizes. The relationship between the RMS focal-spot size and the average of focal-spot sizes determined by the star and slit methods for 32 focal spots is shown in Figure 34. With only three exceptions, the average focal-spot sizes agreed with the RMS focal-spot size within ±0.1 mm, as indicated by the dotted lines; for small focal spots in the range 0.2–0.4 mm, they agreed within ±0.05 mm. The relative standard deviation of the differences, as an estimate of this approximation, was slightly less than 10%. In practice, the focal-spot size obtained by the pinhole method can be used instead of that obtained by the slit method because both methods tend to overestimate focal-spot sizes by a similar amount. There was not much difference in calculated average focal-spot sizes when the pinhole and slit methods were used, although the slit method provided values that were slightly closer to the RMS focal-spot size.

From these results, we conclude that, for the majority of focal spots, the RMS focal-spot size can be represented within ±10% by the average of the star and slit (or pinhole) focal-spot sizes. We believe that this simple

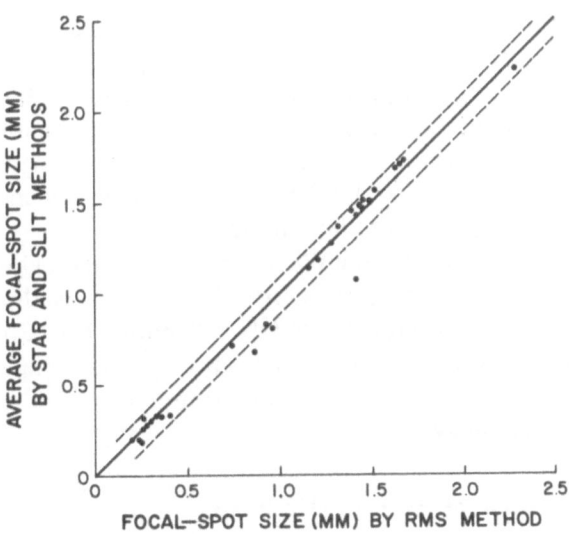

Figure 34. Relationship between focal-spot sizes determined by RMS method and the average focal-spot size determined by star and slit methods.

relationship, if used cautiously, will provide a basis for a practical approach to determining focal-spot sizes.

4. RADIOGRAPHIC NOISE

When a radiographic screen-film system is exposed to X-rays with a uniform intensity distribution, the microscopic density distribution of the developed film fluctuates around the average density. This fluctuation, often called "radiographic noise" or "radiographic mottle," is caused by X-ray quantum-statistical fluctuation, film graininess, and screen structure mottle.[7,9,120] The radiographic noise has been quantified by the Wiener spectrum, which represents the noise components in the spatial-frequency domain in a way similar to the MTF, which was discussed earlier.

Recently, radiographic noise has been recognized as the most important factor affecting the detection and detail visibility of radiologic objects. For example, in high-resolution skeletal radiography,[94] the ultimate limiting factor for detail visibility is the high noise level of the radiographic image, even though skeletal images look almost noiseless when compared to conventional radiographic images made with screen-film systems. In magnification radiography, the effective noise in the rare-earth system is reduced substantially because of the enlargement of the X-ray pattern relative to the recording system, and therefore the detail visibility of small radiologic objects can be improved significantly.[121] In clinical applications, the magnification technique has shown a better radiographic image quality and improved diagnostic accuracy, as demonstrated in mammography[91,122] and cholecystography.[92,123] In addition, recent comprehensive studies[124] on detecting simple objects, such as small plastic spheres, indicated that the measured detectabilities are very closely related to the signal-to-noise ratio calculated from radiographic noise and other imaging parameters based on a statistical decision theory model.[125-127] These findings underscore the need for a better understanding of radiographic noise.

Our knowledge regarding the Wiener spectrum of radiographic mottle is rather limited. One of the reasons for this situation is the difficulty of measuring the Wiener spectrum;[32,33,128-131] another has been the delay in recognizing its importance. The significance of radiographic mottle is increasing steadily as ultrafast, rare-earth screen-film systems are introduced into diagnostic radiology.[132-134]

4.1. Wiener Spectrum of Radiographic Noise

The noise in a radiographic film may be described in terms of fluctuations in density or transmittance. Let us assume that $D(x, y)$ is a point-by-point density distribution obtained with a microdensitometer by two-dimensional scanning. In terms of the average density $\bar{D}$ and the fluctuating component $\Delta D(x, y)$, we can write $D(x, y) = \bar{D} + \Delta D(x, y)$. Similarly, if the transmission is measured with an optical instrument, the transmission distribution, the average, and the fluctuating component are related to $T(x, y) = \bar{T} + \Delta T(x, y)$. Since density and transmission are related as $\bar{D} = -\log \bar{T}$, the fluctuating components can be expressed by

$$\Delta D(x, y) = -\log_{10}\left\{1 + \frac{\Delta T(x, y)}{T}\right\}$$

$$= -(\log_{10}e)\left\{\frac{\Delta T(x, y)}{\bar{T}} - \frac{1}{2}\left(\frac{\Delta T(x, y)}{\bar{T}}\right)^2 + \frac{1}{3}\left(\frac{\Delta T(x, y)}{\bar{T}}\right)^3 - \cdots\right\} \tag{32}$$

Since the relative fluctuation of the transmission is assumed to be small $[\Delta T(x, y)/T \ll 1]$, the higher order terms can be neglected, and Eq. (32) is approximated by

$$\Delta D(x, y) = -0.43\frac{\Delta T(x, y)}{\bar{T}} \tag{33}$$

Therefore, noise in the radiographic film can be evaluated by measuring fluctuations in either denstiy or transmission, and it is possible to convert from one parameter to the other.

The Wiener spectrum $\Phi_D(u, v)$ of radiographic noise is defined in terms of density fluctuations by

$$\Phi_D(u, v) = \lim_{\substack{X \to \infty \\ Y \to \infty}} \overline{\frac{1}{4XY}|F_D(u, v)|^2} \tag{34}$$

where

$$F_D(u, v) = \int_{-X}^{X}\int_{-Y}^{Y} \Delta D(x, y)\exp\{-2\pi i(ux + vy)\}\,dx\,dy \tag{35}$$

The bar indicates the ensemble average. If the transmission is measured, the Wiener spectrum $\Phi_T(u, v)$ is defined by

$$\Phi_T(u, v) = \lim_{\substack{X \to \infty \\ Y \to \infty}} \overline{\frac{1}{4XY}|F_T(u, v)|^2} \tag{36}$$

where

$$F_T(u, v) = \int_{-X}^{X}\int_{-Y}^{Y} \frac{\Delta T(x, y)}{\bar{T}}\exp\{-2\pi i(ux + vy)\}\,dx\,dy \tag{37}$$

Therefore, the relationship between the two Wiener spectra defined in Eqs. (34) and (36) is given by

$$\Phi_D(u, v) = (0.43)^2 \Phi_T(u, v) \tag{38}$$

if the approximation in Eq. (33) is used. The Wiener spectrum has the dimension of an area (mm^2, μm^2, etc.) regardless of whether density or transmission is measured.

It is known that radiographic mottle is composed of quantum mottle, film graininess, and structure mottle.[7, 9] If we assume that noise patterns that are due to individual components are statistically independent of each other, the Wiener spectrum of radiographic mottle $\Phi_R(u, v)$ is given by

$$\Phi_R(u, v) = \Phi_Q(u, v) + \Phi_F(u, v) + \Phi_S(u, v) \tag{39}$$

where Φ_Q, Φ_F, and Φ_S are the Wiener spectra for quantum mottle, film graininess, and structure mottle, respectively. Quantum mottle is usually regarded as dominant in modern screen-film systems.[9, 135] The Wiener spectrum of quantum mottle can be described in a first-order approximation if one assumes that the screen-film system is a transducer for white noise that converts X-ray quanta absorbed in the screen to a visible noise pattern by means of transmission or density fluctuations in the film. Since each absorbed quantum is smeared by the light diffusion in the screen-film system and since the appearance of the noise pattern is influenced by the film contrast, namely, the gradient G of the film, the Wiener spectrum of quantum mottle is given in terms of density by

$$\Phi_{Q,D}(u, v) = (0.43)^2 \frac{G^2}{\bar{n}} \text{MTF}^2 \tag{40}$$

where $\bar{n}$ is the average number of X-ray quanta absorbed by the screen per unit area.[7, 9] If the Wiener spectrum is determined in terms of transmission, we can use Eq. (38) to obtain

$$\Phi_{Q,T}(u, v) = \frac{G^2}{\bar{n}} \text{MTF}^2 \tag{41}$$

Although this model of quantum mottle is very useful in relating imaging parameters to the Wiener spectrum in a simple manner, it is highly idealized. Noise patterns due to quantum mottle are formed through complex processes that include at least (1) the absorption of X-ray quanta at different depths in a phosphor layer;[136] (2) reabsorption of fluorescent X rays[52, 137-141] emitted when the incident X-ray energy is above the K-edge energy of the heavy element of the phosphor; (3) a complicated process of energy conversion from X rays absorbed to light pulses emitted;[142-144] and (4) the broad spectral composition of an incident X-ray beam.

4.2. Measurement of Wiener Spectra

Although the Wiener spectrum is a two-dimensional quantity based on two spatial-frequency axes u and v, its measured values are usually displayed in one-dimensional form. This is because the Wiener spectrum of radiographic noise is usually regarded as being rotationally symmetric. Measuring the Wiener spectrum by scanning methods, such as those described in this chapter, is therefore devised to provide directly only one section of the two-dimensional Wiener spectrum.

The noise sample is scanned by a long, narrow slit in a direction parallel to its narrow width a; the slit length is b. A rectilinear scan by the digital technique (Section 4.2.2) or a circular scan by the analog technique (Section 4.2.1) provides a one-dimensional fluctuating signal that is subjected to digital or analog Fourier analysis. The measured one-dimensional Wiener spectrum $\Phi_m(u)$, which is obtained with this scanning method in terms of transmission, is related to the two-dimensional Wiener spectrum by

$$\Phi_m(u) = \int_{-\infty}^{\infty} \Phi_T(u, v) |\text{MTF}_S(u, v)|^2 \, dv \qquad (42)$$

That is, the scanning operation corresponds to integration along the spatial-frequency axis v, corresponding to the direction of the slit length, and the noise pattern is degraded by the modulation transfer function MTF_S of the microdensitometer.[30, 145] If MTF_S is dominated by the transfer function of the slit, then we have

$$\Phi_m(u) = \int_{-\infty}^{\infty} \Phi_T(u, v) \left| \frac{\sin \pi a u}{\pi a u} \frac{\sin \pi b v}{\pi b v} \right|^2 \, dv \qquad (43)$$

If the significantly nonzero region of the transfer function of the slit length is small compared to the domain of v over which the Wiener spectrum can be considered constant and equal to $\Phi_T(u, 0)$, then Eq. (43) can be written approximately as

$$\Phi_m(u) = \frac{1}{b} \Phi_T(u, 0) \left(\frac{\sin \pi a u}{\pi a u} \right)^2 \qquad (44)$$

Therefore, if the Wiener spectrum is measured by using a narrow long slit only at low spatial frequencies, where $(\sin \pi a u)/\pi a u \simeq 1$, a section of a true two-dimensional Wiener spectrum is given by the product of the slit length and the measured one-dimensional Wiener spectrum; i.e.,

$$\Phi_T(u, 0) = b\Phi_m(u) \qquad (45)$$

This is the basis for using a one-dimensional scanning of the noise pattern to derive a section of the two-dimensional Wiener spectrum.[30, 145]

The actual density or transmission whose fluctuations are being measured depends on the optical instrument used and its optical coupling to the emulsion being evaluated. Densitometers that illuminate the sample with a narrow beam of light and collect transmitted light from 2π sr measure diffuse transmission density.[146] Densitometers that both illuminate with and collect very narrow beams measure specular density, which depends on the scattering properties of the emulsion as well as on the collecting properties of the instrument optics. The ratio of the specular to the diffuse density of a sample is called the Callier coefficient Q of the sample. Most microdensitometers both illuminate and collect light over small angles; the sine of the half-angle ranges from 0.05 to 0.25. (This latter quantity is the numerical aperture (N.A.) of the lens when the light path is through air.) Since the illumination and collection angles are finite in practical micro-densitometers and vary between instruments, the density D_I of a film sample measured with a particular instrument will be characteristic of the instrument used.

If the density fluctuations ΔD in Eq. (34) are measured in terms of D_I, then the resulting Wiener spectrum will have a magnitude that is related to the microdensitometer used for the measurement. It is desirable to convert ΔD to fluctuations of diffuse densities D_D when the magnitude of the Wiener spectrum is compared between laboratories or when the measured Wiener spectrum is combined with the gradient, which is usually determined in terms of diffuse density as discussed in Section 2.3, to estimate the noise equivalent quanta (NEQ) or the detective quantum efficiency (DQE).[147-149]

This conversion can be done by either of two methods. One is a straightforward calibration of the instrument density D_I for diffuse density D_D by scanning a sensitometric strip that contains a series of known diffuse-density levels. When this calibration is not possible, the Wiener spectrum in terms of diffuse density Φ_D can be obtained from the Wiener spectrum in terms of instrument density Φ_I by

$$\Phi_D(u, v) = \Phi_I(u, v)/\dot{Q}_I^2 \tag{46}$$

where $\dot{Q}_I$ $(= \Delta D_I/\Delta D_D)$ is the slope of a characteristic curve, which is a plot of instrument density versus diffuse density, at the diffuse density of the film scanned for the Wiener spectrum measurement. The term $\dot{Q}_I$ is different from the *instantaneous* Q $(= D_I/D_D)$ value. If Q is used instead of $\dot{Q}_I$ for this conversion, the error for XRP film at a density of 1.0 has been estimated to be approximately 13%.[33]

4.2.1. ANALOG METHOD

A device for measuring the Wiener spectrum[33] has been constructed in the Rossmann Laboratory at the University of Chicago and used to study the

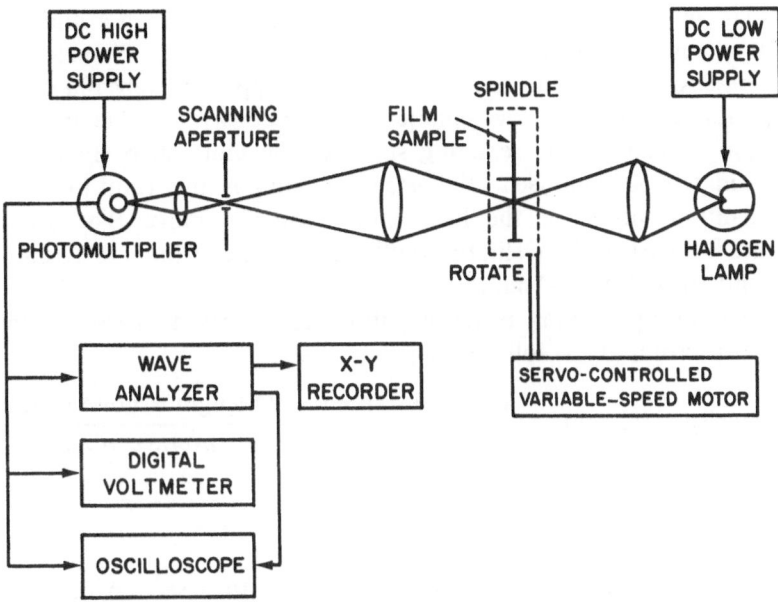

Figure 35. Schematic diagram of analog system for measuring Wiener spectra.

effect of quantum mottle on radiographic image quality. A schematic diagram of the measuring system is shown in Figure 35. The device consists of an optical system similar to that of a microdensitometer with a rotating spindle, a wave analyzer with constant filter bandwidth, and a recorder. The processed-film sample, which was exposed to a uniform beam of X rays, is mounted on the spindle. While the film is being illuminated, the spindle is rotated by the motor, and the transmitted light is collected by a photomultiplier behind an aperture. Therefore, the light spot scans the film circularly, and the transmission fluctuation of the film is converted into a current fluctuation in the photomultiplier. The fluctuating photocurrent is supplied to the wave analyzer and a voltmeter and oscilloscope. The noise power spectrum of the current fluctuation, which is proportional to the squared output of the wave analyzer, is registered on an X-Y recorder by a slow sweeping of the filter's frequency. The digital voltmeter measures the average photocurrent corresponding to the average transmission of the film. The oscilloscope monitors the fluctuating photocurrent and also the output from the wave analyzer.

In the Wiener spectrum analyzer, the following factors are taken into consideration: The stability of the two regulated dc power supplies for the lamp and the photomultiplier is better than one part in 10^5. In addition, a low-noise photomultiplier is used, which keeps the instrument's noise at a low level. The optics applied to the radiographic film consists of a microscope objective lens (10 × /0.30) and an eyepiece lens (10 ×). The diameter of the scanning circle on the film can be varied up to approximately 100 mm, and the rotational speed of the spindle is continuously variable from 5 to 1000 rpm. Therefore, the scanning speed of the film can be increased to approximately 5000 mm/sec. High-speed scanning is important for measuring quantum mottle in the low-spatial-frequency range. The scanning aperture is a narrow, long slit; the width can be varied from 0 to 10 mm and the length, from 0 to 100 mm.

Relationships among physical parameters involved in measuring the Wiener spectrum are as follows:

$$\text{Wiener spectral value} \equiv \frac{1}{2}\frac{\text{scanning speed} \times \text{slit length} \times \text{noise voltage}}{\text{filter band width} \times \text{average voltage}}$$

$$\text{Spatial frequency} \equiv \frac{\text{temporal frequency}}{\text{scanning speed}}$$

The derived Wiener spectral value corresponds to a section of the two-dimensional Wiener spectrum and has the dimension of mm^2. The factor 1/2 is used for obtaining the Wiener spectrum on only one side of the spatial-frequency axis, since the noise measurement by the electronic filter includes components of both positive and negative frequencies. The scanning-speed and the electronic-filter bandwidth are used to convert the electronic noise per unit temporal frequency to spatial noise per unit spatial frequency. The slit length is applied to yield a section of the two-dimensional Wiener spectrum from the one-dimensional Wiener spectrum. The noise voltage is divided by the average voltage because our measurements of the Wiener spectrum are based on the transmission fluctuation relative to the average transmission of the film. The spatial frequency is obtained by dividing the temporal frequency by the scanning speed.

One of the important technical factors affecting the Wiener spectrum measurements by the electronic Fourier analysis method is the choice of the scanning aperture. Several considerations enter into selecting aperture dimensions. The frequency response or the bandwidth of the optical system for measuring the one-dimensional Wiener spectrum depends on the width of the scanning slit. The narrower the slit width, the better the frequency response will be. However, the narrow slit width may require using a high photomultiplier voltage, which tends to increase the instrument noise. As stated before, the slit length has to be very large when a section of the two-dimensional Wiener spectrum is determined.[145] The slit should be long enough so that the Wiener spectrum is flat compared to the MTF associated with the length of the aperture. An exceedingly long slit, however, reduces the current fluctuation relative to the average current of the photomultiplier and thus increases the instrument noise relative to the current fluctuation. The length is in the useful range as long as a logarithmic plot of slit length versus the square of the measured noise voltage is a negative diagonal straight line. If the slit length is too small, the limits of integration in Eq. (43) are determined by $\Phi_T(u, v)$ instead of by $(\sin \pi b v)/\pi b v$, the square of the noise voltage becomes constant, and the curve indicates saturation.

Determining the Wiener spectrum by this method requires taking an ensemble average of the measurements; that is, measurements must be repeated with scans of film samples having the same statistical noise property, and results are then averaged. We are currently averaging four measurements. In addition, the Wiener spectra at frequencies from 0.1 to 7.5 cycles/mm are derived by two separate measurements of two frequency ranges, namely, a low range 0.1–1.5 cycles/mm and a midrange 0.5–7.5 cycles/mm. This is necessary because of the limited useful range of the wave analyzer. The bandwidth of the electronic filter in the wave analyzer is 100 Hz, and a frequency range 100–1500 Hz is employed for Wiener spectral measurements. The fluctuation of the data is generally greater in the low-frequency than in the high-frequency range.

Film samples used for this measurement are approximately 100 mm in diam., and the average gross diffuse density of a sample is usually kept at 1.00 $\pm$ 0.03. The scanning slit used for measuring radiographic mottle has a 1-μm width and 1-mm length in the film plane. We had previously examined the effect of the slit length from 1 to 4 mm for three screen-film systems and found that there was no significant effect of the slit length over this range. Recently, Sandrik and Wagner[150] investigated the role of slit length in determining the Wiener spectra of screen-film systems and found that using a 1-mm slit length may decrease Wiener spectral values at low frequencies by 10–20% in comparison with a *long* slit, which can provide a *true* section of the two-dimensional Wiener spectrum.

The reproducibility of the measurements has been examined in terms of the standard deviation of two independent measurements on the same film sample. The average standard deviation from 0.1 to 7 c/mm was approximately 4%. Usually, uncertainties are greater in the Wiener spectrum data at low spatial frequencies. This is due to increased statistical uncertainties when obtaining low-frequency noise, which forms a coarse, *large* pattern. A large-diameter scanning circle is, therefore, required if this uncertainty is to be reduced.

4.2.2. DIGITAL METHOD

The Wiener spectrum of radiographic mottle can be determined by a digital Fourier transformation of noise data obtained from microdensitometer scans. Since digital computers are now readily available, the digital method has recently become popular at many institutions.[33,151,152] With this method, the Wiener spectrum may be calculated directly by a two-dimensional Fourier transform[152] or a one-dimensional Fourier transform of the scanned data in a way similar to the analog method discussed in Section 4.2.1. Both methods employ a fast Fourier transform (FFT) algorithm; however, using the one-dimensional FFT will be more practical in many institutions where computation time is limited; the two-dimensional FFT may be used by those who have ready access to a large computer.[152]

The digital approach discussed here is based on the method used at the Bureau of Radiological Health (BRH).[33,130,131,153] A film sample is scanned over a central 10-cm square region in a rectilinear raster pattern by a Perkin-Elmer PDS microdensitometer. The scanning slit, which is imaged through a $4\times$, 0.11 N.A. objective and a $10\times$ eyepiece lens, has dimensions of 15×588 μm^2 at the film plane. Measuring a film sample involves a set of many scans, each 10 cm long, with data acquired at 10-μm intervals. After each scan, the film is displaced perpendicular to the scan direction by the effective length of the slit. The light transmitted through the film is detected by a photomultiplier tube. The tube output is passed through a logarithmic amplifier to produce a voltage proportional to the optical density, and this voltage is digitized by a 10-bit analog-to-digital converter.

In order to produce one-dimensional scan data for a one-dimensional FFT, we synthesized the density fluctuation corresponding to a narrow- and long-slit scan which could yield a section of the two-dimensional Wiener spectrum, as discussed in Section 4.2, by averaging transmission values derived from density measurements of adjacent points in a direction perpendicular to the scan. According to Sandrik and Wagner,[150] the slit length required to give accurate Wiener spectral values at low frequencies is quite long. For example, when the Wiener spectra at 0.4 cycle/mm were within 5% of the plateau, slit lengths of at least 4.2, 2.6, and 2.5 mm for Hi-Plus, Par Speed, and Detail screens with XRP film were required.

To provide some protection against the aliasing error, which tends to increase the measured Wiener spectra at high frequencies, high-frequency (low-pass) filtering was applied digitally by averaging two adjacent data points in a direction parallel to the scan. The effect of this high-frequency filtering was corrected later by multiplying the inverse of its MTF with the calculated Wiener spectrum. The same high-frequency filtering can be performed analogously when the noise film sample is scanned with a wide-scanning aperture in the microdensitometer. In fact, a scanning aperture having approximately twice the sampling distance is often used for this purpose.[152]

Low-frequency (high-pass) filtering is applied next in order to eliminate very low-frequency components due to physical defects in the emulsion layer[151] and screens, to roller marks from the film processor, or to nonuniform X-ray intensity distribution. This is a baseline correction where the local density averages are used. If the density scan data at a point j are expressed as D_j, the noise fluctuation ΔD_j which will be subjected to the FFT and windowing is written as

$$\Delta D_j = D_j - \bar{D}_j \tag{47}$$

where the local density average $\bar{D}_j$ is determined from

$$\bar{D}_j = \frac{1}{N} \sum_{K=j-(N/2)}^{j+(N/2)-1} D_K \tag{48}$$

Here, ΔD_j is the deviation of the D_j values from a local mean of $\bar{D}_j$ over N points. A value of $N = 1000$ ($N \Delta X = 0.10$ mm; distance increment $\Delta X = 0.01$ mm) was chosen at the BRH for these measurements because it yielded the best reproducibility of low-frequency results and because of the shape of the low-frequency region of the Wiener spectrum of film graininess.

After filtering, data were selected from 68 overlapping segments of $L\,(=256)$ data points with an overlap of $L/2$. A window was applied to the data within each segment that produced the weighting

$$D_{jw} = \Delta D_j \, W_j \tag{49}$$

where

$$W_j = 1 - \{[j - (L+1)/2]/(L+1)/2\}^2, \qquad 1 \leqslant j \leqslant L \tag{50}$$

Using this window in the domain of the data has approximately the same effect as using the hanning (Tukey) window in the domain of the Wiener spectrum. The spectral bandwidth of this window is approximately 0.45 cycles/mm (FWHM). Since windowing is basically a smoothing technique, the effect on the Wiener spectrum of radiographic mottle may be very small when another smoothing technique is applied in the frequency domain.

The FFT algorithm is now applied to each segment of the windowed data, and the estimated Wiener spectrum $\Phi'(u)$ at the spatial frequency u is given by

$$\Phi'(u) = \frac{L \, \Delta X}{U} \left| \sum_{j=1}^{L} D_{jw} \, e^{-2\pi i m(j-1)/L} \right|^2 \tag{51}$$

where $u = m \, \Delta u$, $\Delta u = 1/L \, \Delta x = 0.39$ cycles/mm, and

$$U = \frac{1}{L} \sum_{j=1}^{L} W_j^2 \tag{52}$$

The spectral estimates are then corrected for the slit width a and the slit length b of the microdensitometer and also for the discrete FFT of the noise data, by

$$\Phi(u) = b \frac{\pi u a}{\sin \pi u a} \frac{\sin \pi u \, \Delta x}{\pi u \, \Delta x} \Phi'(u) \tag{53}$$

When the Wiener spectrum is evaluated in the low-frequency range, as is usually the case for radiographic mottle, the effect of the second and third factors is usually small.

Finally, spectral estimates from 68 segments of each scan and from a total of ten scans are averaged to yield the final Wiener spectrum.

The standard error (S.E.) of the Wiener spectrum components depends on the total length X of the film scanned and the frequency resolution Δu of the measured spectrum, as given by

$$\text{S.E.} = \left(\frac{1}{X \Delta u} \right)^{1/2} \tag{54}$$

At the BRH, when the Wiener spectra are measured with $X = 1000$ mm and $\Delta u = 0.39$ cycle/mm, the predicted standard error of a spectral estimate is about 5%. This value is in good agreement with independent measurements on a given sample. These concepts and the rationale for segmenting the data have been discussed in greater detail.[153]

At the Los Alamos National Laboratory, Hanson et al.[152] has obtained a two-dimensional Wiener spectrum by two-dimensional sampling of noise data and application of two-dimensional FFT. The high-frequency portion of the Wiener spectrum was measured by sampling with a $25 \times 25 \, \mu m^2$ aperture in 12-μm steps in both directions. A total film area of 1.84×1.84 cm^2 was segmented into 256×256 pixel regions. The average Wiener spectrum was obtained for 36 regions. The low-frequency region was determined with a $100 \times 100 \, \mu m^2$ aperture measured in 40-μm steps. A total film area of 7.17×7.17 cm^2 was segmented into 49 regions, each 256×256 pixels in size. The low- and high-frequency Wiener spectra were in close agreement between 2 and 6 cycles/mm.

The Wiener spectra of Hi-Plus/XRP and Detail/XRP measured at Los Alamos agreed closely with those at the BRH.[152] In addition, when the analog method was applied to these film samples, the results were in close agreement with those obtained at the BRH; details of this comparison are reported elsewhere.[33]

4.3. Experimental Results

Measured Wiener spectra for Trimax 12 screens with various films are shown in Figure 36. We prepared noise film samples by using uniform X-ray exposures at 80 kV with a 20-mm aluminum filter. The average density of

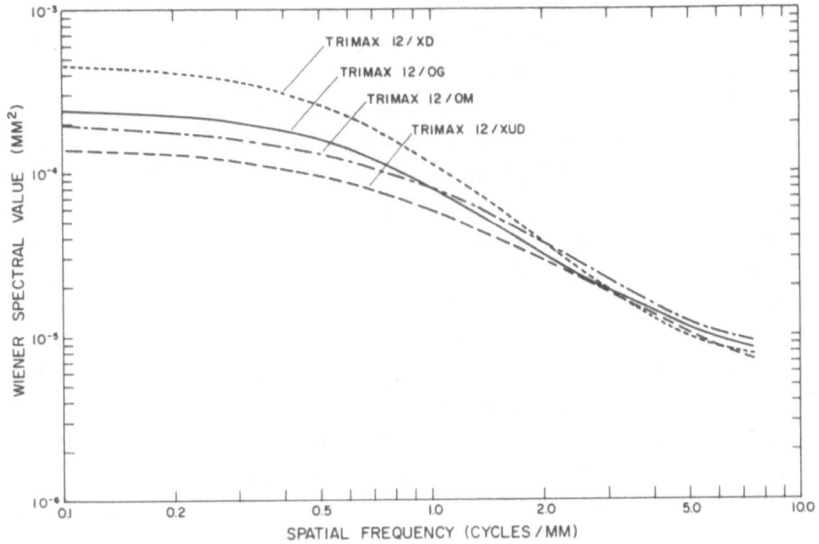

Figure 36. Wiener spectra of Trimax 12 screens with various films.

these films was approximately 1.00. The samples are prepared on different occasions and processed under different conditions. Therefore, we could not use the relative speeds of Trimax 12 screens with OM, XUD, OG, and XD film, which were 2.1, 2.5, 4.1, and 8.4, respectively (Section 3.2.3) in a straightforward manner to estimate relative exposures required to obtain these film samples. These results are shown merely to illustrate the effect of some physical parameters on the measured Wiener spectra.

The Wiener spectra in Figure 36 were measured with the analog method described in Section 4.2.1, and they are plotted in terms of relative transmittance. We repeated the measurements by scanning the same noise sample independently at least twice and then averaged the results. The standard deviation of two independent measurements was approximately 4%, which gives an indication of the reproducibility of the measurements. Recently, we employed a polynomial curve-fitting technique to smooth out the measured Wiener spectra in the frequency domain; this slightly improved our estimate of the Wiener spectra.[124]

Of the systems included in Figure 36, Trimax 12/OM is a single-screen/single-emulsion film system;[51] Trimax 12/XUD is an anticrossover system;[48] Trimax 12/XD is an ultra-high-speed system;[154] and Trimax 12/OG employs a medium-speed green-sensitive film, which may be regarded as a standard system here. The variation in the Wiener spectra presented here may be interpreted by using Eq. (41), which is related to the gradient, the MTF, and the average number of X-ray quanta absorbed per unit area of the screens, as discussed in the following. First, the spatial extent of the Wiener spectrum of quantum mottle is determined by the square of the MTF. Among the screen-film systems examined, Trimax 12 with OM and XUD are high-resolution systems, as discussed in Section 3.2.3. The high level of the MTFs of these two systems is demonstrated by the relative magnitude of the Wiener spectra at high spatial frequencies. The Wiener spectrum of Trimax 12/OM crosses the Wiener spectra of Trimax 12/OG and XD. Second, at low frequencies, where the MTFs of screen-film systems may be considered to be almost 1.0, the Wiener spectrum of quantum mottle is proportional to the square of the gradient and the reciprocal of the average number of X-ray quanta absorbed in the screens. However, when only the film is changed for a given pair of screens, such as Trimax 12 used with XD, OG, and XUD, this relationship can be simplified because the basic X-ray absorption property in the screens is the same. In such a case, the Wiener spectrum of quantum mottle is simply proportional to the film speed and the square of the gradient. This is basically applicable to the comparison of the three Wiener spectra for Trimax 12 with XD, OG, and XUD. The speeds of XD or XUD film are approximately two times or one-half the speed of OG film, respectively, and the gradients of these films are comparable in a first-order approximation. In Figure 36, the low-frequency levels of the Wiener spectra of Trimax 12 with XD or XUD relative to that of Trimax 12/OG are very close to these speed ratios, and therefore Eq. (41) appears to be useful as a guide that relates some parameters to the Wiener spectrum of quantum mottle. Additional examples are described in greater detail elsewhere.[48, 51, 154]

The Wiener spectrum of Trimax 12/OM at low spatial frequencies is lower than that of Trimax 12/OG; this implies that the number of X-ray quanta absorbed in Trimax 12/OM is greater than that for Trimax 12/OG. Even though Trimax 12/OM is a single system and thus inherently less X-ray absorbing, the difference between the two systems is due to the increased incident X-ray exposure of Trimax 12/OM, since the speed of Trimax 12/OM is considerably lower than that of Trimax 12/OG.

In summary, the Wiener spectra of screen-film systems vary considerably even when the same screens are used with different films. This variation is related to the speed, gradient, and MTF of the system. In addition, it has been shown that the Wiener spectrum of a screen-film system changes with the X-ray beam quality used.[129, 138, 139]

ACKNOWLEDGMENTS

The author is greatly indebted to the late Kurt Rossmann for his pioneering work on the analysis of radiographic images and for establishing the facilities for radiologic imaging research at the University of Chicago. Developments, refinements, and improvements of methods and techniques for quantitative analysis of radiographic imaging systems at the Rossmann Laboratory were made by many individuals presently or previously associated with this laboratory; they are Philip A. Bunch, Heang-Ping Chan, David G. Goodenough, Arthur G. Haus, Yoshiharu Higashida, Gunnila Holje, Masamitsu Ishida, Yoshie Kodera, Leh-Nien Loo, Maryellen Lissak Giger, Charles E. Metz, Robert A. Schmidt, Kenneth A. Strubler, and Carl J. Vyborny.

I am grateful to Mrs. E. Lanzl for editing the text and to Miss Evelyn Ruzich for typing the manuscript.

This work was supported by the National Institute of General Medical Sciences through USPHS grant GM 18940, by the National Cancer Institute through USPHS grant CA 24806, and by the Bureau of Radiological Health through USPHS contracts FDA 223-78-6003 and 223-80-6003.

REFERENCES

1. K. Rossmann, Image quality and patient exposure, *Curr. Probl. Radiol.* **2**, 1–34 (1972).
2. K. Doi, K. Rossmann, and A. G. Haus, Image quality and patient exposure in diagnostic radiology, *Phot. Sci. Eng.* **21**, 269–277 (1977).
3. K. Rossmann, Modulation transfer function of radiographic systems using fluorescent screens, *J. Opt. Soc. Am.* **52**, 774–777 (1962).
4. R. H. Morgan, L. M. Bates, U. V. Gopala Rao, and A. Marinaro, Frequency response characteristics of X-ray films and screens, *Am. J. Roentgenol.* **92**, 426–440 (1964).
5. M. Hofert, Messung der Kontrastubertragungsfunktion von Röntgenverstärkerfolien, *Acta Radiol. Diagn.* **1**, 1111–1122 (1963).
6. K. Doi, Measurement of optical transfer functions of intensifying screens, *Oyo Buturi* **33**, 50–52 (1964).
7. K. Rossmann, Spatial fluctuations of X-ray quanta and the recording of radiographic mottle, *Am. J. Roentgenol.* **90**, 863–869 (1963).
8. K. Rossmann, Recording of X-ray quantum fluctuations in radiographs, *J. Opt. Soc. Am.* **52**, 1162–1164 (1962).
9. K. Doi, Wiener spectrum analysis of quantum statistical fluctuation and other noise sources in radiography, in *Television in Diagnostic Radiology* (R. D. Moseley and J. H. Rust, eds.), pp. 313–333, Aesculapius, Birmingham (1969).
10. H. M. Cleare, personal communication.
11. A. G. Haus and K. Rossmann, X-ray sensitometer for screen-film systems used in medical radiography, *Radiology* **94**, 673–678 (1970).
12. J. Hale and P. Block, Step-wedge sensitometry, *Radiology* **128**, 820–821 (1978).
13. *Sensitometric Properties of X-Ray Films*, Radiography Markets Division, Eastman Kodak, Rochester, N.Y. (1979).
14. D. R. Bednarek and S. Rudin, Comparison of modified boot-strap and conventional sensitometry in medical radiography, *Proc. SPIE* **233**, 2–6 (1980).
15. L. K. Wagner, A. G. Haus, G. T. Barnes, J. A. Bencomo, and S. R. Amtey, Comparison of methods used to measure the characteristic curve of radiographic screen/film systems, *Proc. SPIE* **233**, 7–10 (1980).

16. M. V. Yester, G. T. Barnes, and M. A. King, Peak kilovoltage boot-strap sensitometry, *Radiology* **136**, 785–786 (1980).

17. G. T. Barnes, The use of bar-pattern test objects in assessing the resolution of film/screen systems, in *Physics of Medical Imaging: Recording System Measurements and Techniques* (A. G. Haus, ed.), pp. 138–151, American Association of Physicists in Medicine, American Institute of Physics, New York (1979).

18. A. G. Haus, K. Rossmann, C. J. Vyborny, P. B. Hoffer, and K. Doi, Sensitometry in diagnostic radiology, radiation therapy, and nuclear medicine, *J. Appl. Photo. Eng.* **3**, 114–124 (1977).

19. G. Holje and K. Doi, Sensitivity, sensitometry, resolution, and noise characteristics of new radiographic screen-film systems, in *HHS Publication FDA 82-8187*, pp. 39–77, Rockville, Maryland (1982).

20. C. E. K. Mees and T. H. James, *The Theory of the Photographic Process*, pp. 409–436, Macmillan, New York (1966).

21. C. J. Vyborny, The speed of radiographic screen-film systems as a function of X-ray energy and its effect on radiographic contrast, Ph.D. dissertation, University of Chicago, Chicago, Illinois (1976).

22. C. J. Vyborny, H and D curves of screen-film systems: Factors affecting their dependence on X-ray energy, *Med. Phys.* **6**, 39–44 (1979).

23. G. Sanderson and G. J. Johnston, Effect of development temperature changes in an automatic processor, *Proc. SPIE* **70**, 26–32 (1975).

24. H. P. Chan and K. Doi, Determination of radiographic screen-film system characteristic curve and its gradient by use of a curve-smoothing technique, *Med. Phys.* **5**, 443–447 (1978).

25. B. Carnahan, H. A. Luther, and J. O. Wilkes, *Applied Numerical Methods*, p. 129, Wiley, New York (1969).

26. T. R. McCalla, *Introduction to Numerical Methods and FORTRAN IV Programming*, p. 239, Wiley, New York (1967).

27. C. E. Metz and K. Doi, Transfer function analysis of radiographic imaging systems, *Phys. Med. Biol.* **24**, 1079–1106 (1979).

28. W. K. Pratt, *Digital Image Processing*, Wiley, New York (1978).

29. K. G. Beauchamp and C. K. Yuen, *Digital Methods for Signal Analysis*, George Allen and Unwin, London (1979).

30. J. C. Dainty and R. Shaw, *Image Science*, Academic, New York (1974).

31. A. Papoulis, *Probability, Random Variables, and Stochastic Process*, McGraw-Hill, New York (1965).

32. K. Doi and K. Rossmann, Measurements of optical and noise properties of screen-film systems in radiography, *Proc. SPIE* **56**, 45–53 (1975).

33. K. Doi, G. Holje, L. N. Loo, H. P. Chan, J. M. Sandrik, R. J. Jennings, and R. F. Wagner, MTFs and Wiener spectra of radiographic screen-film systems, in *HHS Publication FDA 82-8187*, pp. 1–77 (1982).

34. R. F. Wagner, K. E. Weaver, E. W. Denny, and R. G. Bostrom, Toward a unified view of radiological imaging systems, Part 1: Noiseless images, *Med. Phys.* **1**, 11–24 (1974).

35. K. Doi, Optical transfer functions of the focal spot of X-ray tubes, *Am. J. Roentgenol.* **94**, 712–718 (1965).

36. A. Bouwers, The practical value of contrast transfer in radiology, in *Diagnostic Radiologic Instrumentation: Modulation Transfer Function*, pp. 3–13, C. C. Thomas, Springfield, Illinois (1965).

37. K. Rossmann, Image-forming quality of radiographic screen-film systems: the line spread function, *Am. J. Roentgenol.* **90**, 178–183 (1963).

38. C. H. Dyer and E. L. Criscuolo, Measurement of spatial-frequency response of certain film-screen combinations to 10-MeV X rays, *Materials Evaluation* 631–636 (1966).

39. J. W. Coltman, The specification of imaging properties by response to sine wave input, *J. Opt. Soc. Am.* **44**, 468–471 (1954).
40. G. Lubberts, Some aspects of the square-wave response function of radiographic screen-film systems, *Am. J. Roentgenol.* **106**, 650–654 (1969).
41. R. P. Rossi, W. R. Hendee, and C. R. Ahrens, An evaluation of rare-earth screen/film combinations, *Radiology* **121**, 465–471 (1976).
42. T. W. Ovitt, R. Moore, and K. Amplatz, The evaluation of high-speed screen-film combinations in angiography, *Radiology* **121**, 449–455 (1975).
43. J. P. Weiss, Notes on determining modulation transfer data for X-ray film-screen combinations, in *Image Analysis and Evaluation* (R. Shaw, ed.), pp. 527–531, SPSE (1977).
44. R. A. Schmidt, The use of lead test patterns for MTF measurement of radiographic screen/film systems, M.Sc. thesis, University of Chicago, Chicago, Illinois (1982).
45. S. Sakuma, Y. Ayakawa, Y. Okumura, and Y. Maekoshi, Determination of focal-spot characteristics of microfocus X-ray tubes, *Invest. Radiol.* **4**, 335–339 (1969).
46. G. Sanderson and H. M. Cleare, MTF of screen-film systems: the influence of screens and crossover, *Photo. Sci. Eng.* **18**, 251–253 (1974).
47. G. U. V. Rao and P. P. Fatouros, Evaluation of a new X-ray film with reduced crossover, *Med. Phys.* **6**, 226–228 (1979).
48. K. Doi, L. N. Loo, T. M. Anderson, and P. H. Frank, Effect of crossover exposure on radiographic-image quality of screen-film systems, *Radiology* **139**, 707–714 (1981).
49. P. C. Bunch, Imaging characteristics of screen-film systems using oblique incidence conditions, Ph.D. dissertation, University of Chicago, (1975).
50. R. E. Wayrynen, R. S. Holland, and R. P. Schwenker, Film-screen sharpness in complex-motion tomography, *Invest. Radiol.* **12**, 195–198 (1977).
51. Y. Higashida, P. H. Frank, and K. Doi, Basic imaging properties of single-screen/single-film combinations and their clinical application in diagnostic radiology, *Radiology* **149**, 571–577 (1983).
52. C. J. Vyborny, C. E. Metz, K. Doi, and A. G. Haus, Calculated characteristic X-ray reabsorption in radiographic screens, *J. Appl. Photo. Eng.* **4**, 172–177 (1978).
53. B. A. Arnold and B. E. Bjarngard, The effect of phosphor K X rays on the MTF of rare-earth screens, *Med. Phys.* **6**, 500–503 (1979).
54. G. Lubberts, The line spread function and the modulation transfer function of X-ray fluorescent screen-film systems—Problems with double-coated films, *Am. J. Roentgenol.* **105**, 909–917 (1969).
55. K. Rossmann and G. Sanderson, Validity of the modulation transfer function of radiographic systems measured by the slit method, *Phys. Med. Biol.* **13**, 259–268 (1968).
56. K. Doi, Computer simulation study of screen-film system nonlinearity in fine-detail imaging, *Phys. Med. Biol.* **18**, 863–877 (1973).
57. K. Strubler, K. Doi, and K. Rossmann, Density dependence of the line spread function of screen-film systems, *Phys. Med. Biol.* **18**, 219–225 (1973).
58. K. Rossmann, G. Lubberts, and H. M. Cleare, Measurement of the line spread function of radiographic systems containing fluorescent screens, *J. Opt. Soc. Am.* **54**, 187–190 (1964).
59. G. Sanderson, Erroneous perturbations of the modulation transfer function derived from the line spread function, *Phys. Med. Biol.* **13**, 661–663 (1968).
60. K. Doi, K. Strubler, and K. Rossmann, Truncation errors in calculating the MTF of radiographic screen-film systems from the line spread function, *Phys. Med. Biol.* **17**, 241–250 (1972).
61. C. E. Metz, K. Strubler, and K. Rossmann, Choice of line spread function sampling distance for computing the MTF of radiographic screen-film systems, *Phys. Med. Biol.* **17**, 638–647 (1972).

62. B. A. Arnold, H. Eisenberg, and B. E. Bjarngard, The LSF and MTF of rare-earth oxysulfide intensifying screen, *Radiology* **121**, 473–477 (1976).

63. K. Doi, G. Holje, L. N. Loo, and H. P. Chan, Evaluation of resolution properties of radiographic screen-film systems, *HHS Publication (FDA) 80-8126*, pp. 162–180, Rockville, Maryland (1980).

64. K. Doi and K. Sayanagi, Role of optical transfer function for optimum magnification in enlargement radiography, *Jap. J. Appl. Phys.* **9**, 834–839 (1970).

65. S. Uchida, Fourier analysis of X-ray tube focal-intensity distribution along the beam through optical system, *Oyo Buturi* **34**, 97–107 (1965).

66. H. Kanamori, Optical transfer function of X-ray tube focal-spot sizes for various tube currents, *Jap. J. Appl. Phys.* **4**, 227–228 (1965).

67. E. Takenaka, K. Kinoshita, and R. Nakajima, Modulation transfer function of the intensity distribution of the roentgen focal spot, *Acta Radiol. (Ther)* **7**, 263–272 (1968).

68. G. U. V. Rao and L. M. Bates, The modulation transfer functions of X-ray focal spots, *Phys. Med. Biol.* **14**, 93–106 (1969).

69. K. Kiviniitty, Modulation transfer function of the focal spot of X-ray tubes, *Comm. Phys. Math.* **40**, 9–39 (1970).

70. K. Doi, L. N. Loo, and H. P. Chan, X-ray tube focal-spot sizes: Comprehensive studies of their measurement and effect of measured size in angiography, *Radiology* **144**, 383–393 (1982).

71. A. Elsas, E. Fenner, R. Friedel, and H. Schnitger, Geometrische Unscharfe und Intensitatsverteilung in einen Röntgenaufnahmefeld, *Fortschr. Rontgenstr.* **115**, 822–827 (1971).

72. P. Brubacher and B. M. Moores, The modulation transfer function of the focal spot with a twin-peaked intensity distribution, *Radiology* **107**, 635–640 (1973).

73. A. E. Burgess, Focal spots: I. MTF separability, *Invest. Radiol.* **12**, 36–43 (1977).

74. A. E. Burgess, Focal spots: II. Models, *Invest. Radiol.* **12**, 44–53 (1977).

75. A. E. Burgess, Focal spots: III. Field characteristics, *Invest. Radiol.* **12**, 54–61 (1977).

76. O. Mattsson, Focal-spot variations with exposure data—important factors in daily routine, *Acta. Radiol. (Diagnosis)* **7**, 161–169 (1968).

77. B. M. Moores and W. Roeck, The field characteristics of the focal spot in the radiographic imaging process, *Invest. Radiol.* **8**, 53–57 (1973).

78. D. J. Klein, R. T. Bergeron, and H. Bernstein, Wide-field resolution in radiography, *Radiology* **122**, 811–815 (1977).

79. K. Doi, Field characteristics of geometric unsharpness due to the X-ray tube focal spot, *Med. Phys.* **4**, 15–20 (1977).

80. P. Spiegler and W. C. Breckinridge, Imaging of focal spot by means of the star test pattern, *Radiology* **102**, 679–684 (1972).

81. H. Bernstein, R. T. Bergeron, and D. J. Klein, The relationship of the radiant intensity distribution of focal spots to resolution, *Radiology* **111**, 427–431 (1974).

82. A. E. Burgess, Interpretation of star test pattern images, *Med. Phys.* **4**, 1–8 (1977).

83. A. E. Burgess, Effect of asymmetric focal spots in angiography, *Med. Phys.* **4**, 21–25 (1977).

84. J. E. Gray, M. P. Capp, and F. R. Whitehead, An improved technique for X-ray image evaluation, *Invest. Radiol.* **9**, 252–261 (1974).

85. K. Doi, B. Fromes, and K. Rossmann, New device for accurate measurement of the X-ray intensity distribution of X-ray tube focal spots, *Med. Phys.* **2**, 268–273 (1975).

86. G. Groh, E. Klotz, and H. Weiss, Simple and fast method for the presentation of the two-dimensional modulation transfer function of X-ray systems, *Appl. Opt.* **12**, 1693–1697 (1973).

87. K. Doi, Advantages of magnification radiography, in *Breast Carcinoma: The Radiologist's Expanded Role* (W. W. Logan, ed.), pp. 83–92, Wiley, New York (1977).

88. K. Doi, H. K. Genant, and K. Rossmann, Comparison of image quality obtained with

optical and radiographic magnification techniques in fine-detail skeletal radiography: effect of object thickness, *Radiology* **118**, 189–195 (1976).

89. M. Pfeiler and K. Dietz, Microfocus X-ray tubes and the image quality obtained with geometric enlargement—comments on future developments of X-ray tubes, in *Small Vessel Angiography* (S. K. Hilal, S. Baum, J. J. Bookstein, R. H. Greenspan, M. P. Judkins, E. J. Potshen, K. Rossmann, and H. H. Ter-Pogossian, eds.), pp. 36–47, C. V. Mosby, St. Louis (1973).

90. H. Eisenberg, M. Braun, B. Arnold, W. Holland, and R. Gould, Evaluation of microfocal-spot X-ray tubes and rare-earth oxysulfide intensifying screens in magnification radiography, *Proc. SPIE* **56**, 184–190 (1976).

91. E. A. Sickles, K. Doi, and H. K. Genant, Magnification film mammography: studies of image quality and clinical evaluation, *Radiology* **125**, 69–76 (1977).

92. H. Imhof and K. Doi, Application of radiographic magnification technique with an ultra-high-speed rare-earth screen-film system to oral cholecystography, *Radiology* **129**, 173–178 (1978).

93. A. G. Haus, K. Doi, J. R. Chiles, K. Rossmann, and R. A. Mintzer, The effect of geometric and recording system unsharpness in mammography, *Invest. Radiol.* **10**, 43–52 (1975).

94. H. K. Genant and K. Doi, High-resolution skeletal radiography: image quality and clinical applications, *Curr. Probl. Diag. Radiol.* **7**, 1–62 (1978).

95. International Commission of Radiological Units and Measurements, *Methods of Evaluating Radiological Equipment and Materials:* recommendations of the ICRU, NBS Handbook 89, GPO Washington, D.C. (1962).

96. National Electrical Manufacturers' Association, *Measurement of dimensions of focal spots of diagnostic X-ray tubes*, NEMA Standard 9-11-1974, Publication No. XR5-1974, New York.

97. A. H. G. Kuntke, On the determination of roentgen tube focal-spot sizes by pinhole camera roentgenography, *Acta. Radiol.* **47**, 55–64 (1957).

98. B. A. Arnold, B. E. Bjarngard, and J. C. Klopping, A modified pinhole camera method for investigation of X-ray tube focal spots, *Phys. Med. Biol.* **18**, 540–549 (1973).

99. International Electrical Commission: Characteristic of focal spots in diagnostic X-ray tube assemblies for medical use, draft proposal (1980).

100. E. Zieler and H. Pulvermacher, Evaluation of size and MTF measurements on X-ray sources, presented at the 14th International Congress of Radiology, October 23–29, Rio de Janeiro (1977).

101. K. Doi and K. Rossmann, Evaluation of focal-spot distribution by RMS value and its effect on blood vessel imaging in angiography, *Proc. SPIE* **47**, 207–213 (1975).

102. M. Braun, W. Roeck, and G. Gillian, X-ray tube performance characteristics and their effect on radiologic-image quality, *Proc. SPIE* **152**, 94–103 (1978).

103. M. Braun, Focal spots in the future of mammography, in *Reduced-Dose Mammography* (W. W. Logan and E. P. Muntz, eds.), pp. 195–209, Masson, New York (1979).

104. G. U. V. Rao and L. M. Bates, Effective dimensions of roentgen tube focal spots based on measurements of the modulation transfer function, *Acta. Radiol. (Ther)* **9**, 362–368 (1970).

105. J. J. Bookstein and W. Steck, Effective focal-spot size, *Radiology* **98**, 31–33 (1971).

106. G. U. V. Rao, A new method to determine the focal-spot size of X-ray tubes, *Am. J. Roentgenol.* **111**, 628–633 (1971).

107. P. J. Friedman and R. H. Greenspan, Observation on magnification radiography: Visualization of small blood vessels and determination of focal-spot size, *Radiology* **92**, 549–557 (1969).

108. Nuclear Associates, Inc., Carle Place, New York, Catalog G-1.

109. S. C. Prasad, W. R. Hendee, and P. L. Carlson, Intensity distribution, modulation transfer function, and the effective dimension of a line-focus X-ray focal spot, *Med. Phys.* **3**, 217–223 (1976).

110. S. C. Prasad and W. R. Hendee, Effective size of the transverse dimension of X-ray tube focal spots, *Med. Phys.* **4**, 235–238 (1977).

111. K. Doi and K. Rossmann, Computer simulation of small blood vessel imaging in magnification radiography, in *Small Vessel Angiography* (S. K. Hilal, S. Baum, J. J. Bookstein, M. P. Judkins, E. J. Potshen, K. Rossmann, and H. H. Ter-Pogossian, eds.), pp. 6–12, C. V. Mosby, St. Louis (1973).

112. S. Wende, E. Zieler, and N. Nakayama, *Cerebral Magnification Angiography*, Springer-Verlag, Berlin (1974).

113. T. Sandor, D. F. Adams, P. G. Herman, H. Eisenberg, and H. L. Abrams, The potential of magnification angiography, *Am. J. Roentgenol.* **120**, 916–921 (1974).

114. K. Doi and K. Rossmann, Effect of focal-spot distribution on blood vessel imaging in magnification angiography, *Radiology* **114**, 435–441 (1975).

115. K. Doi, L. N. Loo, and K. Rossmann, Validity of computer simulation of blood vessel imaging in angiography, *Med. Phys.* **4**, 400–403 (1977).

116. E. N. C. Milne, Characterizing focal-spot performance, *Radiology* **111**, 483–486 (1974).

117. R. T. Bergeron, Manufacturers' designation of diagnostic X-ray tube focal-spot size: a time for candor, *Radiology* **111**, 487–488 (1974).

118. H. Bernstein, R. T. Bergeron, and D. J. Klein, Routine evaluation of focal spots, *Radiology* **111**, 421–425 (1974).

119. W. R. Hendee and E. L. Chaney, X-ray focal spots: practical considerations, *Appl. Radiol.* **3**, 25–29 (1974).

120. G. T. Barnes, Radiographic mottle: A comprehensive theory, *Med. Phys.* **9**, 656–667 (1982).

121. K. Doi and H. Imhof, Noise reduction by radiographic magnification, *Radiology* **122**, 479–487 (1977).

122. E. A. Sickles, Microfocal-spot magnification mammography using xeroradiographic and screen-film recording systems, *Radiology* **131**, 599–607 (1979).

123. K. Rossmann, J. R. Williams, and D. J. Goodenough, Evaluation of radiologic-image quality, *Proc. SPIE* **35**, 75–81 (1973).

124. L. N. Loo, Correlation between visual- and physical-image quality indices: Detectability of nylon bead images in radiographic noise, Ph.D. dissertation, University of Chicago (1982).

125. R. F. Wagner, Decision theory and the detail signal-to-noise ratio of Otto Schade, *Photo. Sci. Eng.* **22**, 41–46 (1978).

126. A. E. Burgess, R. F. Wagner, and R. J. Jennings, Human signal detection performance for noisy medical images, *Proc. IEEE* (to be published).

127. M. Ishida, K. Doi, L. N. Loo, C. E. Metz, and J. L. Lehr, Digital-imaging processing: effect on the detectabilities of simulated low-contrast radiographic patterns, *Radiology* **150**, 569–575 (1984).

128. R. F. Wagner, Toward a unified view of radiological-imaging systems. Part II: Noisy images, *Med. Phys.* **4**, 279–296 (1977).

129. G. T. Barnes, The dependence of radiographic mottle on beam quality, *Am. J. Roentgenol.* **127**, 819–824 (1976).

130. R. F. Wagner and K. E. Weaver, Noise measurements on rare-earth intensifying screen systems, *Proc. SPIE* **56**, 198–207 (1975).

131. R. F. Wagner, Fast Fourier digital quantum mottle analysis with application to rare-earth intensifying screen systems, *Med. Phys.* **4**, 157–162 (1977).

132. R. A. Buchanan, S. I. Finkelstein, and K. A. Wickersheim, X-ray exposure reduction using rare-earth oxysulfide intensifying screens, *Radiology* **105**, 185–190 (1972).

133. R. F. Wagner and K. E. Weaver, Prospects for X-ray exposure reduction using rare-earth intensifying screens, *Radiology* **118**, 183–188 (1976).

134. A. L. N. Stevels, New phosphors for X-ray screens, *Med. Mundi* **20**, 12–22 (1975).
135. H. M. Cleare, H. R. Splettstossor, and H. F. Seemann, An experimental study of the mottle produced by X-ray intensifying screens, *Am. J. Roentgenol.* **188**, 168–174 (1962).
136. G. Lubberts, Random noise produced by X-ray fluorescent screens, *J. Opt. Soc. Am.* **58**, 1475–1483 (1968).
137. R. E. Shuping and P. F. Judy, Energy absorbed in calcium tungstate X-ray screens, *Med. Phys.* **4**, 239–243 (1977).
138. C. J. Vyborny, L. N. Loo, and K. Doi, The energy-dependent behavior of noise Wiener spectra in their low-frequency limits: comparison with simple theory, *Radiology* **144**, 619–622 (1982).
139. G. Holje, An investigation of imaging properties of radiographic screen-film systems, Ph.D. dissertation, the University of Lund, Lund, Sweden (1983).
140. H. P. Chan and K. Doi, Energy and angular dependence of X-ray absorption in screen-film system, *Phys. Med. Biol.* **28**, 565–579 (1983).
141. C. E. Metz and C. J. Vyborny, Wiener spectral effects of spatial correlation between the sites of characteristic X-ray emission and reabsorption in radiographic screen-film systems, *Phys. Med. Biol.* **28**, 547–564 (1983).
142. R. K. Swank, Absorption and noise in X-ray phosphors, *J. Appl. Phys.* **44**, 4199–4203 (1973).
143. C. E. Dick and J. W. Motz, Image information transfer properties of X-ray fluorescent screens, *Med. Phys.* **8**, 337–346 (1981).
144. C. E. Dick, J. W. Motz, and H. Roehrig, New method for the experimental determination of the detective quantum efficiency of X-ray screens, *Proc. SPIE* **233**, 11–15 (1980).
145. K. Doi, Scans in measuring Wiener spectra for photographic granularity, *Jap. J. Appl. Phys.* **5**, 1213–1216 (1966).
146. J. H. Altman, Sensitometry of black-and-white materials, in *The Theory of the Photographic Process*, 4th ed. (T. H. James, ed.), p. 181, MacMillan, New York (1977).
147. R. C. Jones, Quantum efficiency of detectors for visible and infrared radiation, in *Advances in Electronics and Electron Products*, XI (L. Marton, ed.), pp. 83–183. Academic, New York (1959).
148. R. Shaw, Evaluating the efficiency of imaging processes, *Rep. Prog. Phys.* **41**, 1103–1115 (1978).
149. J. M. Sandrik and R. F. Wagner, Absolute measures of physical-image quality: measurement and application to radiographic magnification, *Med. Phys.* **9**, 540–549 (1982).
150. J. M. Sandrik and R. F. Wagner, Radiographic screen-film noise power spectrum: variation with microdensitometer slit length, *Appl. Opt.* **20**, 2795–2798 (1981).
151. M. DeBelder and J. DeKerf, The determination of the Wiener spectrum of photographic emulsion layers with digital methods, *Photo. Sci. Eng.* **11**, 373–378 (1967).
152. J. M. Sandrik, R. F. Wagner, and K. E. Hanson, Radiographic screen-film noise power spectrum: calibration and intercomparison, *Appl. Opt.* **21**, 3597–3602 (1982).
153. R. F. Wagner and J. M. Sandrik, An introduction to digital noise analysis, in *The Physics of Medical Images: Recording System Measurements and Techniques*, pp. 524–545 (A. G. Haus, ed.), American Institute of Physics, New York (1979).
154. K. Doi, Y. Kodera, L. N. Loo, H. P. Chan, and Y. Higashida, MTFs and Wiener spectra of radiographic screen-film system, vol. II, *HHS Publication (FDA)* (to be published).
155. C. J. Vyborny, C. E. Metz, and K. Doi, Large-area contrast prediction in screen-film systems, *Proc. SPIE* **23**, 30–36 (1980).
156. H. P. Chan, Investigation of physical characteristics of scattered radiation and performance of antiscatter grids in diagnostic radiology: Monte Carlo simulation studies, Ph.D. dissertation, University of Chicago (1981).
157. H. P. Chan and K. Doi, Physical characteristics of scattered radiation and performance of antiscattered grids in diagnostic radiology, *RadioGraphics* **2**, 378–406 (1982).

158. H. P. Chan and K. Doi, The validity of Monte Carlo simulation studies of scattered radiation in diagnostic radiology, *Phys. Med. Biol.* **28**, 109–129 (1983).
159. J. A. Sorenson and J. A. Nelson, Investigation of moving-slit radiography, *Radiology* **120**, 705–711 (1976).
160. G. T. Barnes, I. A. Brezovich, and D. M. Witten, Scanning multiple-slit assembly: A practical and efficient device to reduce scatter, *Am. J. Roentgenol.* **129**, 497–501 (1977).
161. H. P. Chan and K. Doi, Investigation of performance of antiscatter grids: Monte Carlo simulation studies, *Phys. Med. Biol.* **27**, 785–803 (1982).
162. K. Doi, P. H. Frank, H. P. Chan, C. J. Vyborny, S. Makino, N. Iida, and M. Carlin, Physical and clinical evaluation of new high-strip-density radiographic grids, *Radiology* **147**, 575–582 (1983).

Index

249